Diagnostic Studies of Microwave Plasmas Containing Hydrocarbons
Using Tunable Diode Lasers

Inauguraldissertation

zur

Erlangung des akademischen Grades

doctor rerum naturalium (Dr. rer. nat.)

an der Mathematisch-Naturwissenschaftlichen Fakultät

der

Ernst-Moritz-Arndt-Universität Greifswald

vorgelegt von

Lars Mechold

geboren am 26.04.1970

in Pößneck

Greifswald, 05.07.2000

Die Deutsche Bibliothek – CIP-Einheitsaufnahme

Mechold, Lars:
Diagnostic studies of microwave plasmas containing hydrocarbons
using tunable diode lasers / vorgelegt von Lars Mechold. - Berlin :
Logos-Verl., 2001

Zugl.: Greifswald, Univ., Diss., 2000
ISBN 3-89722-687-1

ISBN 3-89722-687-1

Logos Verlag Berlin
Comeniushof, Gubener Str. 47,
10243 Berlin
Tel.: +49 030 42 85 10 90
Fax: +49 030 42 85 10 92
INTERNET: http://www.logos-verlag.de

Dekan: Prof. Dr. K. Fesser

1. Gutachter: Prof. Dr. J. P. F. Conrads (Universität Greifswald)

2. Gutachter: Dr. P. B. Davies (University of Cambridge)

Tag der Promotion: 12. Dezember 2000

Kurzfassung

Seit mehr als einhundert Jahren werden Plasmen technologisch eingesetzt. Auch wenn oftmals im Verborgenen, so hat die Anwendung von Plasmaprozessen unseren Alltag doch nachhaltig beeinflußt. Die Entwicklung der Elektronik hin zur Computertechnik und deren explosionsartige Verbreitung in den letzten zwanzig Jahren ist ohne neuartige plasmagestützte Herstellungsprozesse nicht vorstellbar. Auch die effektivere Lichterzeugung in bewährten oder neuartigen Strahlungsquellen sowie der Ersatz gesundheitlich bedenklicher, aber bis heute unverzichtbarer Substanzen stehen im Mittelpunkt der Plasmaanwendungen. Die im Plasma vorkommenden Prozesse sind sehr vielfältig und im Allgemeinen miteinander verkoppelt. Bildungs- und Vernichtungsphänomene stehen im Gleichgewicht und bestimmen im Wesentlichen das besondere plasmatypische Gemisch aus Elektronen, Photonen, Ionen und Neutralteilchen. Festkörperoberflächen, die mit diesem Ensemble in Verbindung stehen, unterliegen einer intensiven Wechselwirkung mit den Teilchen und der Strahlung des Plasmas.

Als Bestandteile vieler gegenwärtiger und zukunftsträchtiger Technologien werden oftmals Niederdruckplasmen im Bereich von 10^{-3} bis 10 mbar eingesetzt. Auch im Hinblick auf die Wahl der Reaktorgeometrie, der Anordnung der Elektroden und der Dimension des Plasmas bieten die Niederdruckplasmen vielfältige Möglichkeiten. Zu ihrer Erzeugung kann man Gleich- oder Wechselspannung zwischen einigen Hz bis GHz verwenden. Die für die Nutzung derartiger Plasmen entscheidenden inneren Eigenschaften, wie Teilchendichte oder Energieverteilung, variieren über weite Bereiche. Der Anteil der Neutralen übersteigt den der geladenen Teilchen um mehrere Größenordnungen.

Nachdem in den Anfängen der Plasmaphysik hauptsächlich Edelgasplasmen untersucht wurden, stehen heute vielfach Molekülgasplasmen im Mittelpunkt technologischer Anwendungen. Es lassen sich Reinigungs- und Ätzprozesse, Schichtbildungen zum Korrosionsschutz oder auch zur gezielten Funktionalisierung von Oberflächen realisieren. Denkt man weiterhin an die plasmagestützte Stoffumwandlung von niedermolekularen Kohlenwasserstoffen in Mitteldestillate oder die Wasserstoffproduktion für eine Brennstoffzelle, so versteht sich leicht die ökonomische Bedeutung solcher Prozesse.

Bisher wurden viele verschiedene Entladungstypen auf ihre Nutzbarkeit hin untersucht, einige von ihnen auch bis zur industriellen Reife gebracht. Im Hinblick auf substratschonende chemisch aktivierte Oberflächenmodifizierung besitzen insbesondere molekulare Mikrowellenentladungen ein großes Potential. Die in diesen Plasmen auftretenden Stoff- und Oberflächenumwandlungen sind durch hohe Dissoziationsgrade der Ausgangsstoffe und einen großen Anteil von chemisch aktiven Radikalen im Plasmavolumen gekennzeichnet.

Unter den vielfältigen Elementarprozessen besitzt die Plasmachemie der Neutralteilchen eine Schlüsselfunktion. Mehr und mehr setzt sich diese Erkenntnis durch und manifestiert sich in zunehmend detaillierteren Modellierungen der möglichen chemischen Reaktionen, sowie Messungen *einzelner* neutraler Spezies in diesen Plasmen. Die hauptsächlich elektronisch induzierten Plasmaprozesse leiten aber ganze Ketten verschiedener chemischer Sekundärreaktionen ein, die die *gesamte* Stoffgruppe der Ausgangsmoleküle umfassen. Fast immer steht der Anwender einer kaum überschaubaren Anzahl von chemischen Verbindungen gegenüber. Darunter befinden sich nicht nur stabile Moleküle sondern insbesondere auch neutrale Radikale, die für den Reaktionsverlauf entscheidende Zwischenprodukte darstellen. Die Notwendigkeit eines tieferen Verständnisses dieser plasmachemischen Prozesse als Voraussetzung einer anwendungsorientierten Optimierung verlangt die Identifizierung der gebildeten Moleküle und nachfolgend deren Quantifizierung.

Daraus leitet sich direkt die Zielstellung der vorliegenden Arbeit ab. Mit Hilfe einer leistungsfähigen Moleküldiagnostik sollte die sich ausbildende chemische Vielfalt in kohlenwasserstoffhaltigen Mikrowellenplasmen untersucht werden. Dazu zählen neben den stabilen Spezies auch die wesentlich reaktiveren Radikale. In den durchgeführten Untersuchungen an Mikrowellenplasmen wurden Prozeßgase verschiedener Mischungsverhältnisse von Wasserstoff zu Sauerstoff neben kleinen Zumischungen von Argon und den Kohlenwasserstoffen Methan oder Methanol verwendet. Unter den veränderlichen Prozeßgaszusammensetzungen waren die Dissoziationsgrade der Ausgangsstoffe gleichermaßen von Interesse wie die Konzentrationen der Zwischen- und Endprodukte. Einen Schwerpunkt der Untersuchungen bildeten die Unterschiede zwischen einem durchströmten und abgeschlossenen Reakor beim Übergang von wasserstoff- zu sauerstoffdominierten Plasmen.

Als Moleküldiagnostik kam die hochauflösende Infrarot-Absorptionsspektroskopie mittels durchstimmbarer Diodenlaser (TDLAS) zum Einsatz. Durch hohe Selektivität und Empfindlichkeit erfüllt TDLAS in kaum vergleichbarer Art und Weise die Forderung nach Konzentrationsmessungen von stabilen und transienten Molekülen. Mit Hilfe der TDLAS wird die Majorität der Plasmateilchen, die neutralen Moleküle in ihrem Grundzustand, unter Plasmabedingungen, in-situ, experimentell zugänglich. Vor der Anwendung als plasmadiagnostische Methode zeigte die TDLAS bereits im Bereich der Umweltanalytik ihr Potential. So konnte durch die Einführung dieser Diagnostik das Verständnis des chemischen Zyklus' der Chlorverbindungen und dessen Wirkung auf das Ozon in der Atmosphäre verbessert werden. Die Kenntnis solcher klimarelevanter Prozesse führte zu weitreichenden Konsequenzen in der Umweltpolitik. Bei der Erforschung atmosphärenchemischer Phänomene zeigte sich die zentrale Bedeutung von neutralen Radikalen. Einen vergleichbaren Stellenwert nehmen Radikale auch in molekularen Plasmen ein.

Ausgangspunkt der durchgeführten Untersuchungen war die Bestimmung der Absolutkonzentration des Methylradikals. Daran schloß sich erstmalig in ihrer Gesamtheit die Konzentrationsmessung der stabilen Moleküle CH_4, CH_3OH, C_2H_2, C_2H_4, C_2H_6, CH_2O, $HCOOH$, CO, CO_2 und H_2O an. Dafür kamen H_2-O_2-Ar Plasmen mit Zumischungen zwischen 0,9 % und 7,2 % von Methan oder Methanol zum Einsatz. Unter der Verwendung von verschiedenen Mischungsverhältnissen von Wasserstoff zu

Sauerstoff bei zunächst konstanten Anteilen von Argon und dem Kohlenwasserstoff wurden die Konzentrationsverläufe jeder einzelnen Spezies aufgenommen. Diese Meßserien wurden für jeweils vier verschiedene Kohlenwasserstoffanteile wiederholt. Alle Experimente wurden bei konstantem Entladungsdruck, Mikrowellenleistung und Gesamtgasfluß durchgeführt. Um Erkenntnisse über die Endprodukte der chemichen Vorgänge zu erhalten, wurden zusätzlich Messungen am abgeschlossen Plasmareaktor durchgeführt. Dabei konnten Effekte durch den aktiven Gasfluß vermieden und so die chemischen Reaktionspfade besser untersucht werden. Während in der Vergangenheit in methanhaltigen Mikrowellenplasmen schon einzelne neutrale Kohlenwasserstoffe vermutet und gefunden wurden sind erstmalig Messungen von Formaldehyd in methanhaltigen Plasmen sowie Methan, Formaldehyd und Ameisensäure in methanolhaltigen Mikrowellenplasmen gelungen.

Unterstützt durch eine experimentbegleitende Modellierung, die neben einer Vielzahl von chemischen Reaktionen insbesondere primäre Elektronenstoßprozesse berücksichtigte, wurde der Versuch unternommen, einen Fortschritt im Verständnis komplexer plasmachemischer Phänomene von kohlenwasserstoffhaltigen Nichtgleichgewichtsplasmen zu erzielen. Die Modellrechnungen basieren auf einem System von Ratengleichungen, das mit Hilfe eines kommerziellen Softwarepakets gelöst wurde. Ausgehend von verschiedenen Quellgasmischungen im abgeschlossen Reaktor wurden die Konzentrationen der verschiedenen beteiligten Spezies zeitlich relaxiert berechnet. Diese Prozedur wurde zunächst auf das sauerstofffreie H_2-Ar-CH_4 Plasma angewendet und später auch auf das O_2-Ar-CH_4 Plasma erweitert. Der Vergleich der theoretischen und experimentellen Ergebnisse zeigt zufriedenstellende Übereinstimmung. Insbesondere fand sich auch in den Modellrechnungen die zentrale Stellung des Methylradikals bezüglich der eingeschlagenen Reaktionspfade wieder.

Die für die Modellierung verwendeten chemischen Gleichungen und die experimentellen Ergebnisse in den H_2-O_2-Ar-CH_3OH Plasmen führten zu einem neuartigen Ansatz für methanolhaltige Plasmen. Aufgrund des auch für diese Plasmen gemessenen umfangreichen Satzes von Molekülkonzentrationen konnten mögliche Reaktionspfade benannt werden.

Im Ausblick wird auf weitere verschiedene Einsatzmöglichkeiten der TDLAS hingewiesen. Dabei lassen sich in Zukunft auch zeitaufgelöste Konzentrationsmessungen realisieren, die das große Potential für weitere Untersuchungen im Bereich der Grundlagenforschung sowie auch für den Einsatz als Prozeßdiagnostik im industriellen Umfeld verdeutlichen.

1 Introduction

Plasmas have been used for technical applications for more than a hundred years. Even if they exist latently the utilisation of plasma processes has influenced the daylife lastingly. Developments from electronics to computer techniques, and then widespread distribution over the last twenty years, would have been impossible without powerful plasma based production processes. Examples like efficient light generation inside long-lifetime radiation sources as well as the replacement of substances which are dangerous to health, but still essential for the process, are of central interest. Many basic physical reactions appear inside the plasma and generally they are coupled with each other. Production and loss phenomena are balanced and determine the mixture of electrons, ions, neutrals and photons in typical plasmas. Solid surfaces contacting the ensemble suffer intense interaction with the plasma particles and the plasma radiation.

Low pressure plasmas in the range of 10^{-3} to 10 mbar are often part of key technologies. Depending on reactor geometry, electrode position and plasma size there are different applications of low pressure plasmas. Beside dc excited plasmas the excitation frequency of ac plasmas varies between a few Hz and GHz. These plasmas are characterised by various properties of specific plasma particles, like their nature, the concentration and the energy distribution. The concentration of neutrals exceeds that of the charged particles by orders of magnitude.

While in the early experiments mainly rare gas plasmas were investigated today molecular plasmas are of growing interest for basic research and technological applications. They are used in cleaning and etching processes as well as for thin film deposition for corrosion protection or specific functionality. Other technologies like conversion of natural gas to liquids or hydrogen production for the fuel cell of a car are examples of the importance of plasma processes. Different types of plasmas have been examined for these purposes and also some of them have found application in industrial technologies.

The molecular microwave discharge is a useful tool for chemically activated surface modification of sensitive substrates. The electron density is typically higher and the mean electron energy is lower than in rf plasmas. If high temperatures at the substrate surface must be avoided or even living materials shall be used the microwave discharge is preferably applied. The electrodeless configuration of this discharge type is advantageous for very clean conditions and provides the possibility of up-scaling. Various technical solutions have been found already to design relatively large discharge dimensions. Therefore the microwave plasma has a high potential of application in plasma technology.

In low pressure microwave plasmas the species and surface conversions are determined by high degrees of dissociation of the precursor molecules and high amounts of chemically active radicals inside the plasma volume. Plasma chemistry plays a key role in the diverse elementary processes. This fact is now gaining more and more acceptance so that an increasing number of more detailed modelling and measurement studies of *single* neutral species inside plasmas are being undertaken. The mainly electron induced plasma reactions lead to *entire* series of different chemical secondary reactions involving the whole group of substances making up the source gas molecules. Not only stable molecules but especially transient species like neutral radicals are often important intermediates. Although the identification of the different compounds is a challenge only the measurement of absolute species concentrations will permit the detailed understanding of the plasma processes, one of the main tasks of the present work.

Specific properties of tunable infrared diode laser absorption spectroscopy (TDLAS), like high selectivity and sensitivity, advantageously lend themselves concentration measurement of stable and transient molecular species. TDLAS has been already used in the field of environmental science. The application of this diagnostic method contributed to improved understanding of the chemical reactions of chlorine compounds with atmospheric ozone. Knowledge of the chemical paths existing in the upper atmosphere relevant to the global climate has lead to far-reaching consequences for political decision making. The basic contribution of neutral radicals to chemical reaction pathways was found. Radicals inside plasmas are of a similar great importance.

Main subject of this work were diagnostic studies of microwave plasmas containing hydrocarbons using TDLAS. For the first time systematic investigations of absolute concentrations of stable and transient carbon-containing species in low pressure microwave plasmas have been performed.

TDLAS has been used:

(i) to detect the methyl radical and ten stable molecules in hydrogen-argon-oxygen microwave plasmas containing a small percentage of methane or methanol, under both flowing and static conditions,

(ii) to determine the degrees of dissociation of the source gas hydrocarbons in various gas mixtures, including the identification of the intermediates and the final products,

(iii) to measure the methyl radical concentration and the concentrations of the stable C-2 hydrocarbons C_2H_2, C_2H_4 and C_2H_6, produced in the plasma when increasing amounts of O_2 were added at fixed methane or methanol partial pressures,

(iv) to detect for the first time formaldehyde, formic acid and methane in methanol microwave plasmas, and formaldehyde in methane microwave plasmas,

(v) in addition to monitor the major products CO, CO_2 and H_2O.

The experiments were also accompanied by model calculations:

(i) to help to predict the concentrations in methane plasmas in the absence of oxygen and the trends for the major chemical product species as oxygen was added,

(ii) and to investigate the main reaction paths by analysing the production and loss reaction rates of every single species. This resulted in a plausible reaction scheme for H_2-O_2-Ar-CH_4 plasmas.

Knowledge of the chemical kinetics of H_2-O_2-Ar-CH_4 plasmas and experimental data on H_2-O_2-Ar-CH_3OH plasmas were used to find a new approach to understand methanol-containing plasmas. An extensive set of measured molecular concentrations was analysed to find a plausible reaction scheme.

At the end of this work an outlook is given to describe the most recent developments of the TDL technique and its potential for further fundamental research as well as for plasma process controlling in industry.

2 Molecular Plasma Diagnostics

The concentration of neutrals in low-temperature low-pressure plasmas is usually some orders of magnitude higher than the concentrations of charged particles. The diagnostic methods described below have been chosen for the determination of the neutral gas composition, mainly hydrocarbons inside microwave discharges.

2.1 Intrusive methods

These methods extract particles from the plasma volume disturbing the local plasma properties of the original volume.

Mass Spectrometry (MS). Mass spectrometry is a widespread diagnostic technique for molecular plasmas. A mass spectrometer consists typically of an ion source, the electromagnetic filter and the ion detector. Neutral molecules from the plasma chamber must diffuse to the ion source via an extraction bleed. After ionisation the ions are filtered and detected. It is possible to determine the whole gas composition in one sample. The spatial resolution is restricted to the environment around the bleed zone inside the plasma chamber. Acceptable time resolutions of the order of seconds or some tenths of seconds can be achieved if measurements are restricted to single mass numbers. The hydrocarbon composition in different rf plasmas [1,2,3] and microwave plasmas [4,5] has been measured. In contrast to the determination of the partial pressures of the stable plasma products the determination of partial pressures of free radicals is a more difficult problem. Radicals can appear as a real plasma product or as a fragmentation product inside the mass spectrometer. A distinction beween the two has been found by optimising the ionisation process. The threshold energy of ionisation is a characteristic value used for determining different radicals in rf methane plasmas [6,7]. In addition photoionisation has been applied for determining single radicals, like the methyl radical [8]. With respect to the neutral gas composition the main drawback of MS is that it only detects ions. On the other hand the plasma ions and their velocity distribution can be investigated directly. MS is a useful diagnostic method for particle flux measurements to the wall [9] and for systematic investigation of the ionic components in a plasma [10,11].

Gas Chromatography (GC). Another method of molecular diagnostics, also involving extracting gas from the volume under investigation, is gas chromatography. It is applied for separating mixures of species and is based on the species distribution between two unmiscible phases. This method has a sensitivity of ~ ppm for detecting neutral molecules, e.g. the stable hydrocarbons during deposition of carbon layers in a rf plasma [12] or hot filament diamond deposition [13]. GC allows the qualitative and quantitative analysis of fairly complex and chemically similar molecular mixtures. Since the identification of molecules by GC is sometimes difficult it is often combined with other methods like MS or absorption techniques. Due to pressure differences between the plasma vessel and the diagnostic device the application to low pressure plasmas needs additional corrections. No radicals can be detected by GC techniques. Nevertheless GC is a standard method in the pharmaceutical and chemical industries [14].

2.2   Non-intrusive methods

All the other methods mentioned here are spectroscopic diagnostic techniques which are generally non-intrusive. Spectroscopic diagnostics can provide information about atom, molecule and ion densities in excited and ground states, as well as about gas, rotational and vibrational temperatures.

Optical Emission Spectroscopy (OES). In low temperature molecular plasmas a variety of processes such as electron collisions, ion-molecule reactions, charge exchange and chemical processes lead to a large number of atoms, molecules, radicals and ions in excited states. The generation of photons is mainly governed by the de-excitation of these species. Provided the influence of self absorption inside the plasma is relatively small or can be neglected (referred to as optically thin plasma conditions) the emitted photons can be detected outside the plasma. Even if self absorption does occur the spectral distribution of the emitted light carries with it information on the plasma conditions. It contains information about the different kinds of plasma species, their number density and temperature, as well as about the strengths of internal or external fields.

A plasma diagnostic technique based on emission spectroscopy has the characteristics of an inverse problem. Usually, integral intensities of emission lines in the line of sight are measured within a certain solid angle with a selected spectral resolution. Only if the plasma is homogeneous over the solid angle being investigated can the local value of the intensity be determined. Otherwise theoretical inversion methods have to be used such as Abel inversion in the case of cylindrical symmetry, or tomography. The measured, spatially localised line emission intensities allow the calculation of population densities of the electronic or, for molecules, rovibronic levels of the species of interest provided transition probabilities are known and the spectrometer functions have been calibrated. The calculation of species densities in the *ground* state from measured line intensities is generally an inverse problem. It requires a theoretical model for the excitation and de-excitation processes, all necessary cross-sections, transition probabilities etc. as well as knowledge about the electron energy distribution function [15]. OES has been used for molecular diagnostics e.g. in diamond deposition microwave plasmas [16,17,18,19].

Laser-Induced Fluorescence (LIF). A variety of experiments has been performed in recent years in the field of laser diagnostics of gas discharges. In LIF measurements atoms, ions or molecules are excited from one energy level to a higher energy level by laser radiation tuned to the transition wavelength. Provided the collisional relaxation of the population established in this way are negligible, then the concentration of atoms in the lower level may be deduced from the subsequent fluorescence provided the medium is optically thin. Such measurements provide a high degree of spatial and temporal (millimetres and nanoseconds) resolution and are the basis of techniques for measuring a range of plasma properties. Very important radicals have been detected, such as the hydroxyl radical in methanol plasmas [20], as well as atomic oxygen [21] and hydrogen [18,22].

Coherent Anti-Stokes Raman Scattering (CARS). The application of CARS has a younger history than LIF. This diagnostic method bases on inelastic Raman-scattering of photons on a molecule. Three incoming laser beams are used, two pump lasers with the same frequency (for experimental simplicity mostly one pump laser is used twice) and one laser with the Stokes frequency of the molecule of interest. The frequency difference of the pump laser and Stokes-laser is tuned to the Raman-resonance of the molecule of interest. Further scattering of the third laser photon is used to shift the anti-Stokes wave to shorter wavelengths. This CARS-signal is emitted as a laser-like beam from the interaction volume due to the phase matching condition. The cross sections of these processes are fairly low so that high power lasers are essential parts of the experimental arrangement.

This method is of increasing interest because of the measurable molecular properties as concentration, rovibrational population distribution and temperature. Its spatial resolution is determined by the destination volume of the crossing laser beams. Because of the grazing incidence it can obtain some mm's. The time resolution is about 10 ns. Molecular concentrations to about 10^{14} cm^{-3} can be determined, e.g. for N_2 [23]. Measurements in the reactive zone of a test reactor for hydrocarbon synthesis have also been reported [24,25]. Recently, the methyl radical and methane concentrations were measured in a microwave plasma with a spatial resolution of 2 mm [26]. The methyl radical concentration was found to be in the range of 10^{12} to 10^{13} cm^{-3}. The application of CARS is mainly restricted to the major plasma species. Several related techniques like Resonance Enhanced CARS (RECARS) or Degenerate Four Wave Mixing (DFWM) have been used to measure molecular concentrations of about 10^{12} cm^{-3}, e.g. the CH radical using RECARS [18,27].

Absorption Spectroscopy (AS). The methods of absorption spectroscopy are of great importance in plasma diagnostics because they provide a means of determining the population densities of species in both ground and excited states. The spectral line positions provide species identification while line profiles are often connected with gas temperature and relative intensities provide information about population densities. An important advantage of AS over OES methods is that only relative intensities need to be measured to determine absolute concentrations, avoiding the problems of complete instrument calibration inherent in the OES methods. Absorption spectroscopy has been applied right across the spectrum from the VUV to the far infrared (FIR). Continuous lamps (e.g. the Xe-lamp for the VIS and NIR, and the D_2-lamp for the UV) and tunable narrow-band light sources (e.g. tunable dye lasers, diode lasers) can be used as external light sources. Two principle cases need to be distinguished, the measurement of the absorption of light emitted by: (i) the plasma itself (*self absorption* or *reabsorption*), or (ii) an external light source. If the former is the case the light from the plasma is measured with and without a retro-reflector or compared with the light emitted directly from an identical plasma. This ensures identical line profiles for the emitted and absorbed light. This method is particularly important for checking the optical thickness of the plasma.

The change in light intensity dI_v when a beam passes through a homogeneous plasma layer of thickness dl is given by the net balance of the intensity due to absorption and emission within the layer dl:

$$dI_v = \left(\varepsilon(v) - \kappa(v)I_v\right)dl \qquad (2.1)$$

where $\varepsilon(v)$ is the emission coefficient and $\kappa(v)$ the absorption coefficient per unit length. The absorption coefficient $\kappa(v)$, describing light absorption in an infinitesimally thin layer of a plasma, is given by

$$\kappa(v) = \sum_i N_i \sigma_i(v) \qquad (2.2)$$

where a summation over all absorbing species and states is included. N_i are the species population densities and $\sigma_i(v)$ their absorption cross sections at frequency v. The line profile P_v is normalised by :

$$\int_{line} f(v - v_0) dv = 1 \qquad (2.3)$$

Assuming the upper level k is not populated, then the absolute value of $\kappa(v)$ results from:

$$\int_{line} \kappa(v) dv = \frac{hv}{c_0} N_i B_{ik} \qquad (2.4)$$

The line absorption coefficient is then given by,

$$\kappa(v) = \frac{hv}{c_0} N_i(\alpha, \beta, \gamma, ...) B_{ik} f(v - v_0) \qquad (2.5)$$

where $N_i(\alpha, \beta, \gamma, ...)$ is the population density of the ith level, which depends on different plasma parameters $(\alpha, \beta, \gamma, ...)$, and B_{ik} is the Einstein coefficient for the transition between levels i and k.

In the case where an external light source has much higher intensity than that of the plasma itself, the absorption of radiation can be described by the Beer-Lambert law which is,

$$I_v(l) = I_v(0) \exp(-\kappa(v) l) \qquad (2.6)$$

$I_v(0)$ and $I_v(l)$ are the fluxes of the radiation entering and leaving the plasma, l is the length of the absorbing (homogeneous) plasma column. This condition is illustrated in Figure 2.1.

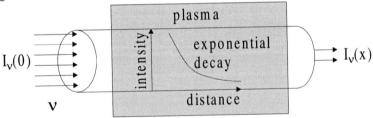

Figure 2.1   Absorption of external radiation in a plasma (Beer-Lambert- Law)

In the further context the absorption coefficient is defined as

$$\kappa(v) = S f(v - v_0) \qquad (2.7)$$

16

where S is the line strength of a transition. This value can also been taken from literature.

It should be noted that in plasma spectroscopy the definition of the effective length of the plasma column is not a trivial question [28]. Inhomogeneity can lead to serious errors. AS only provides integrated intensities along the line of sight [15,18].

Cavity Ring Down Spectroscopy (CRDS). The direct absorption technique CDRS has been applied for 12 years [29]. The basic principle is the measurement of the decay rate of a laser pulse inside an optical cavity formed by two highly reflective mirrors. The trapped light passes through the volume under investigation inside this cavity many times giving extremely long effective absorption lengths, up to several kilometres. The maximum decay time can reach values of 100 μs. Different rates of absorption can be found for the empty cavity and the cavity with a gas sample. The concentration of interest may be deduced from the decay rate. CRDS has proven to be a valuable addition to detection techniques like LIF or simple AS in order to improve the sensitivity by several orders of magnitude. Because of the multireflection in the optical cavity the spatial resolution is limited to the total cavity volume. CRDS has been used for density measurements of SiH_2 radicals and of nanometer size dust particles in silane plasmas [30], and for the detection of N_2^+ ions in a nitrogen discharge [31]. Further this technique has been used for measurements of hydroxyl and methyl radicals in flames and discharges [32,33]. A good survey of using this new technique was given by Engeln [34]. In spite of the increasing application of CRDS there is a major drawback connected with the optical cavity. Deposits inside molecular plasmas strongly reduce the reflectance of the mirrors and weaken the optical cavity. Further the reflectance of the mirrors depends on the wavelength so that for different regions different mirrors are needed. Actually a pair of mirrors works only for a relatively narrow wavelength window. Moreover, in the infrared region the lower reflectance of the mirrors compared to visible or ultraviolet radiation makes its application more difficult. The application of infrared CRDS e.g. for detection of ethylene in the 10-11 μm region, has been reported in ref. [35].

Other absorption techniques have been applied because many of the, however, molecules used and their decomposition products are infrared-active.

Fourier Transform Infrared Spectroscopy (FTIR). This method is a widely used absorption technique using a continuous broad band radiation source and wavelength calibration by interferometric techniques. The wavelength region 2 - 25 μm can be covered and modern spectrometres also extend to the visible range. In contrast to dispersion techniques the FTIR spectrometer records the whole spectrum simultaneously. In principle the resolution can be as high as 0.002 cm^{-1}, determined by the distance scanned by the moveable mirror, but at the expense of time resolution. Fractional absorption as small as 10^{-4} can be measured under optimum conditions. Because of the fast development of computers the mathematical part of Fourier transformation included in FTIR is no longer a problem. All these features have lead to highly integrated spectrometers which are commercially available. Beside species identification information about the molecular structure is also available. It is possible to determine absolute species concentrations of stable molecular species, as well as vibrational and rotational temperatures. FTIR spectroscopy has been used for *in situ* studies of molecular plasmas for a number of years [36]. In the case of plasmas containing high amounts of different infrared active molecules the absorption bands may overlap with even higher vibrational or rotational bands which appear close to the ground state bands. Also small dilutions of water or carbon dioxide somewhere in the optical path make difficulties in molecule identification thus in many cases these spectral positions have to be excluded. This method is generally insufficiently sensitive for detecting free radicals or ions in processing plasmas [37].

Tunable Diode Laser Absorption Spectroscopy (TDLAS). With the development of tunable narrow-band light sources, e.g. tunable dye lasers and infrared diode lasers, continuous sources have been substituted for numerous absorption experiments. These narrow-band light sources have the advantage of high spectral intensity, small bandwidth, and the capability of tuning the radiation over the absorption profile. For a large number of molecular species infrared absorption spectroscopy with tunable diode lasers in the spectral region between 3 and 20 μm is a modern, promising technique. TDLAS has been applied as a plasma diagnostic technique in hydrocarbon-containing plasmas for only 10 years [38,39]. It is mainly used for measuring number densities of stable molecules and especially radicals. This method is characterised by high selectivity, due to the spectral resolution of 10^{-4} cm^{-1}, and absorbances of 10^{-4}. With sufficient line strengths a sensitivity of 10^{10} molecules cm^{-3} can be reached for many molecules. The detection of neutral ground state molecules allows the direct monitoring of the major species in the plasmas. Spatial resolution can be achieved by reducing the cross section of the beam to several mm and by moving it with an optical system. The time resolution of the absolute concentration measurement can basically be as small as ms. It can also be used to measure neutral gas temperatures [40] and to investigate the dissociation processes in different molecular low-temperature plasmas [41,42,43,44].

TDLAS is the method of choice to investigate plasma chemistry in low-pressure planar microwave discharges using a hydrogen-oxygen-argon source gas mixture with small amounts of methane or methanol. A key objective of this type of study is the ability to detect as many plasma species as possible. Therefore, TDLAS is the most versatile technique to measure absolute concentrations of different radicals and stable molecules because it has a low detection limit.

A description of the discharge and diagnostic methods used here can be found in chapters 3 and 4, respectively.

2.3   Methods for the methyl radical

The hypothesis that the methyl free radical is the most likely precursor for formation of carbon films is now gaining acceptance, hence the great importance of suitable diagnostic methods for its detection. Two different possibilities can be distinguished, optical methods and mass spectroscopy. The latter diagnostic techniques must be generally improved to the level of Threshold Ionisation Mass Spectrometry (TIMS) or Photo Ionisation Mass Spectrometry (PIMS) for detecting radicals. TIMS uses the direct ionisation threshold of 9.8 eV for the neutral methyl radical. Measurements of the methyl radical using TIMS were published by several authors [6,7,45]. In addition to investigations of the methyl radical in the plasma volume particle flux measurements to the wall have also been carried out by TIMS [46]. In PIMS a UV radiation beam is used for ionisation of the methyl radical. Measurements using PIMS were published by Celii and Butler [47] and Ando et al. [8]. These modern measurements represent a combination of mass spectroscopy and radiative interaction.

The electronic spectrum of the methyl radical was first detected in the vacuum ultraviolet by Herzberg and Shoosmith in 1956, and later analysed by Herzberg [48,49]. Over one decade later Tan, Winer and Pimentel produced CH_3 by flash photolysis of CH_3I and recorded its low resolution IR absorption spectrum using a dispersion spectrometer [50]. In these pioneering IR experiments in which a carbon arc light source was used the sensitivity was, not surprisingly, rather low. Much higher sensitivity and resolution was achieved later by Hirota and co-workers using infrared lasers, in particular TDLAS [51,52,53]. This greater sensitivity enabled the reaction kinetics of the methyl recombination reaction and of the reaction of methyl radical with oxygen to be investigated, using laser photolysis of CH_3I to generate CH_3 [54].

While the methyl radical is still the subject of basic research in recent years it has been widely studied in applied investigations [55,56,57,58,59,60]. In 1984 Tachibana et al measured the rate of dissociation of CH_4 in an RF methane plasma by TDLAS and predicted that CH_3 should be the most abundant radical in the plasma [61]. Later Celii and co-workers detected the absorption spectrum of the methyl radical in a hot filament CVD reactor using TDLAS and made rough estimates of the concentrations of CH_3, C_2H_2, CH_4, and C_2H_4 near the filament [62]. Recently concentration measurements of the methyl radical by means of CARS in a microwave process plasma with a spatial resolution of two millimetres were published [26].

Most of the measurement techniques for detecting the methyl radical are based on absorption spectroscopy either with 216 nm ultraviolet radiation, e.g. number density measurements by Menningen, Child et al. [63,64], or in the infrared near 606 cm^{-1}. The infrared TDLAS technique has proven to be the most useful because it can also measure the concentrations of related species provided they are IR active. Wormhoudt demonstrated this flexibility by measuring CH_3 and C_2H_2 in a CH_4/H_2 RF plasma using a long path plasma absorption cell [38]. A highly important study for quantifying the concentrations of methyl radicals was the determination of the line strength of the $Q(8,8)$ line of the methyl radical at 608.3 cm^{-1} by Wormhoudt and McCurdy [65]. Systematic TDLAS measurements of several different hydrocarbons including methyl in a 20 kHz methane plasma in a parallel plate reactor were reported by Davies and Martineau [39,41,66]. Goto and co-workers have published numerous studies of methyl and methanol concentrations in rf and ECR plasmas under different conditions e.g. investigating the influence of rare gases on the plasma. They have also combined IR absorption with emission spectroscopy, and investigated the effect of water vapour on the methyl radical concentration in argon/methane and argon/methanol RF plasmas using TDLAS [42,67,68,69,70,71,72]. Kim et al measured CH_3, C_2H_2 and CH_3OH concentrations in methanol/water RF discharges by TDLAS and found that methanol was almost completely dissociated even at medium applied power [73].

3 The Microwave Discharge

Discharges excited by oscillating electric fields are commonly used over a wide range of high frequencies for technological purposes. The specific combination of charged particles and neutral background gas, due to the exciting and ionising interaction between the particles under the influence of the oscillating electric field, is useful for many modern applications. Typical chemical vapour deposition (CVD) devices used for etching or deposition as well as for surface activation and controlled deactivation mechanisms employ high frequency discharges. These plasma sources, especially in industrial environments, are regarded as a „Black Box". To improve the understanding of the plasma itself detailed investigations are necessary.

Microwave plasmas can be excited without electrodes in the reaction chamber. This opens up possibilities for very clean conditions and also for extending the size of a plasma. With molecular source gases this type of plasma is characterised by high degrees of dissociation of the source gas species and high chemical reactivity due to the large concentrations of transient and stable chemically active neutral species present. All applied frequencies of this part of the electromagnetic spectrum must be legally permitted [74]. Frequencies higher than 300 MHz are referred to microwave frequencies. A frequency of 2.45 GHz is one of the most often used for industrial applications, e.g. for diamond deposition [75]. All experiments described in this thesis were done with this microwave frequency.

3.1 Power transfer behaviour

Plasmas are composed of different types of particles, namely electrons, ions and neutrals. For the power level commonly used only charged particles can be influenced by an external electric field. Due to respective masses only the electrons are accelerated significantly and pick up velocity in the electric field.

The elastic collisions of the electrons with the background particles, mainly neutrals, under the influence of the harmonic electric field can be described by a simplified approach [74]. In case the thermal motion of electrons is neglected the equation of motion of a single electron is

$$m_e \frac{d\vec{v}_T}{dt} = -e\vec{E}(t) - m_e v_c \vec{v}_T \qquad (3.8)$$

where m_e is the electron mass and v_c is an average electron-neutral collision frequency for momentum transfer. The first term on the right hand side of (3.8) describes the force exerted by the electric field and the second term represents the „frictional" force due to collisions with the heavy particles. Introducing the steady-state and the complex form of the velocity vector[1] yields for the equation of motion

---

[1] $\vec{v}_T = \text{Re}(\vec{v}_{Tc} \exp i\omega t)$, where $\vec{v}_{Tc}$ is the complex amplitude of the velocity.

$$i\omega m_e \vec{v}_{Tc} = -e\vec{E} - m_e v_c \vec{v}_{Tc} \qquad (3.9)$$

The solution for the complex amplitude of the velocity is

$$\vec{v}_{Tc} = -\frac{e\vec{E}}{m_e}\frac{1}{v_c + i\omega} \qquad (3.10)$$

In general, Maxwell's equations describe the transmission of electromagnetic waves within a medium, and also in the plasma. The electron current density can be calculated using equation (3.10) and yields

$$\vec{J} = -en_e \vec{v}_{Tc} = \sigma\, \vec{E} \qquad (3.11)$$

where n_e is the electron number density. The permittivity σ and the permeability ε of the plasma follow:

$$\sigma = \frac{n_e e^2}{m_e}\frac{1}{v_c + i\omega} \qquad (3.12)$$

$$\varepsilon = \varepsilon_r + i\varepsilon_i = 1 - i\frac{\sigma}{\omega\varepsilon_0} = 1 - \left(\frac{\omega_p}{\omega}\right)\frac{1}{1 - i\frac{v_c}{\omega}} \qquad (3.13)$$

where ω_p is the plasma frequency[2]. The identification of material properties like the permittivity and the permeability makes clear the similarity of plasmas to other states of matter, like solids, liquids or gases. However, all the material constants of the plasma depend on internal parameters like the electron number density or the electron collision frequency. The external properties can be influenced by internal fluctuations. The transmission of microwaves for small losses is important. Three cases can be distinguished:

a) $\omega < v_c$; the value of σ is a real number and the plasma is a conductive medium. Microwave power can be absorbed.

b) $v_c < \omega < \omega_p$; is a transition region. The plasma is still conductive and it absorbs microwave power very well.

c) $\omega > \omega_p$; the plasma is a dielectric medium.

The limiting value, the so called cut-off frequency is reached in the case where the excitation frequency is equal to the plasma frequency. One can also find a corresponding electron number density. It describes the precise conditions when a plasma switches to a dielectric medium. Nevertheless power coupling is best close to the cut-off frequency.

In the ideal case of no collisions between the electrons and heavy particles the plasma behaves like a medium without losses. The permittivity σ is an imaginary number and the permeability ε is real. Only ordered oscillations of the electrons occur because of the external electric field. On average, there is no energy transfer between the

---

[2] Electron plasma frequency: $\omega_p = \sqrt{\dfrac{e_0^2 n_e}{\varepsilon_0 m_e}}$

electromagnetic field and the electrons or the heavy particles. The plasma is an entirely passive medium.

Collisions between the electrons and other particles introduces a disorder in their motion. The electrons are deflected on every collision. This opens up a channel by which energy can flow from the external electric field to the whole ensemble. Because of the large mass difference between the light and heavy particles the electrons will pick up energy from the external field and the electrons will also transmit energy to the heavy particles.

At first only elastic collisions are assumed. The average kinetic energy per period picked up in three-dimensional space is

$$\bar{u} = \frac{e^2 \hat{E}^2}{2m_e \left(v_c^2 + \omega^2 \right)} \tag{3.14}$$

where $\hat{E}^2 = 0.5EE^*$. E^* denotes the conjugate complex. This value of the kinetic energy is much smaller than for ionisation processes of the neutrals. A rough estimate will be given as illustration. A typical value of the field amplitude is 50 Vcm^{-1} and the collision frequency is typically on the order of 10^{10} s^{-1}. Using this value yields a kinetic energy of 1.6 x 10^{-2} eV per period, which is more than two orders of magnitude smaller than characteristic ionisation energies.

Although the electrons are disturbed by every collision even so it is possible for them to follow every field oscillation. However, they are able to pick up energy continuously from the external field, due to the elastic collisions. The mean power transferred per electron Θ_A is

$$\Theta_A = -e\overline{\vec{E}(t)\vec{v}_{Tc}(t)} = \frac{e}{m_e} \frac{v_c \hat{E}^2}{v_c^2 + \omega^2} = 2v_c \bar{u} \tag{3.15}$$

In the absence of collisions with the background gas the electron velocity vector is perpendicular to the field direction so no power is transferred from the electric field to the electron. Including such collisions the random energy of the electron rises continuously. At specific energies excitation or ionisation processes occur and the electron transmits its energy by an inelastic collision. The electron looses most of its energy and then is accelerated again by the external electric field [74].

Above it is considered the behaviour of only one electron. It is necessary to take into account all electrons, leading to the definition of the electron energy distribution function (EEDF). This approach also includes highly energetic electrons, which are assumed to play an important role in such discharges. Furthermore the EEDF is also connected with internal plasma parameters.

As was shown earlier the dependence of power input on the excitation frequency is essential for the microwave discharge. Not only this frequency itself but also the ratio between the excitation frequency and characteristic plasma parameters, like the electron-neutrals collision frequency for the momentum transfer or the plasma frequency, strongly influence the plasma properties. For the same power density a higher electron density has been found for microwave discharges than that of other rf discharges[76].

The electron energy decreases with increasing frequency. The population density of higher energetic levels including the ionised levels and number densities of reactive species is higher in microwave discharges. Several scientific contributions have been published about the dependence of the EEDF on the excitation frequency [74,76,77]. It has been shown how the EEDF changes with the field within a quasi-stationary regime. At low frequencies the electrons are still able to follow the electric field oscillation. The power introduced in equation (3.15) depends on the energy and the field frequency. The maximum of $v_c/(v_c^2 + \omega^2)$ occurs at $v_c = \omega$ and this is the reason why collisions with frequencies close to the excitation frequency are very efficient. Such collision frequencies are again strongly dependent on the gas or the gas mixture. In the case when the electron-electron collisions and the collisions of the electrons with excited species are neglected the EEDF depends only on the ratios $\hat{E}/N$ or ω/N. With increasing field frequency the electrons and thus also the EEDF are no longer able to follow the external field behaviour and a mean value will appear for the losses. The EEDF becomes independent of time and of $\hat{E}/N$. The shape of the EEDF depends on the parameter v_c/ω. For values $v_c/\omega > 1$ the distribution function changes quasi-stationary within the oscillating field. In this situation the electron behaviour is very similar to that in a dc field of a positive column in a glow discharge. When $v_c/\omega >\approx 1$ the EEDF is not Maxwellian but is close to it and for values $v_c/\omega < 1$ the energy of the electrons is a Maxwellian distributed [77].

3.2 Planar discharges

Microwave plasmas can be excited in several different ways. The common feature of all of them is the electrode-free coupling of the electromagnetic energy into the plasma chamber. Only a few typical examples will be mentioned here, e.g. surface wave discharges, resonator discharges and electron cyclotron resonance (ECR) discharges. The latter type of microwave discharge needs in addition a magnetic field.

Non-resonant microwave structures with travelling wavefronts without any external magnetic field are the type of planar microwave plasma sources used in the present studies. This planar discharge was expected to be useful for applications requiring microwave power distributed over a fairly large surface [78].

3.2.1 Construction fundamentals

The phenomena in the planar microwave plasma source can be explained using the principles of the wave propagation in a hollow waveguide at very high frequencies. The object of interest is not the theory of wave guiding but some of the characteristics of the parts used in construction. The microwaves are guided to the plasma source by hollow waveguides. A schematic of the waveguide used with a side ratio of a:b = 2:1, and the characteristic patterns of wall currents are shown in Figure 3.2.

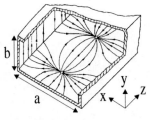

Figure 3.2   *Waveguide co-ordinates and the wall current distribution of the H_{10}-mode [79].*

The current flow on the wide side (a) forms a complex pattern while that on the small side is directed perpendicular to the propagation direction of the microwave.

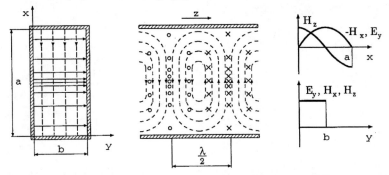

Figure 3.3   *Field distributions of the H_{10}-mode in a rectangular waveguide (solid line: electric field lines; dashed line: magnetic field lines) [80]*

The typical field structure of the H_{10}-mode in a rectangular waveguide is illustrated in Figure 3.3. It can be seen that all field parameters and therefore also the electric field only depends on the x and z components, and not on the y component. This fact leads to the use of this type of waveguide for microwave transmission for plasma generation. Electromagnetic waves are absorbed by the plasma. The actual practical arrangement of the reactor is shown in Figure 3.4.

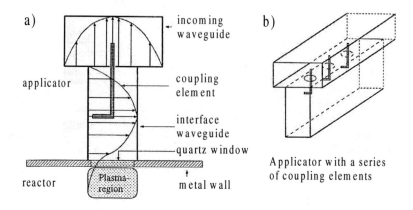

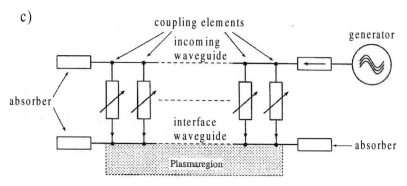

Figure 3.4   *Working acpects of the planar microwave applicator [78,81](a - cross section of the waveguide with a coupling element; b - sequence of coupling elements for generation of extended plasma columns; c - technical scheme)*

The source consists of two waveguides which are placed in a T-shape (Figure 3.4). One end of the upper waveguide is connected to the microwave source and all other ends of the waveguide are terminated by matching absorbers. This configuration only allows travelling microwave modes because the absorbers are designed as ideal sinks for the microwave energy. The discrete adjustable coupling elements are used for the gradual coupling of microwave energy from the upper to the lower interface waveguide. The lower metal wall of the interface waveguide is replaced by a quartz window which is transparent for microwaves. This geometrical and homogeneous interface permits power absorption and plasma generation underneath the quartz window as the electric field penetrates the reaction chamber. The plasma itself forms the fourth wall of the lower waveguide. The settings of the coupling elements allows the local power input to be varied. The size of the waveguide limits the width of the plasma column, but plasma columns of large length can be arranged. Larger width are possible when many waveguides are packed side by side [82].

3.2.2 Properties of the planar discharge

All types of electrical discharges are mainly determined by the external electrical field. The ideal field strength distribution in the planar microwave plasma source is illustrated in Figure 3.5.

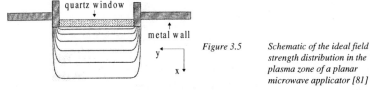

Figure 3.5   *Schematic of the ideal field strength distribution in the plasma zone of a planar microwave applicator [81]*

The electrical field in the plasma column is assumed to be very similar to that of the waveguide itself. The field strength in the y-direction should be constant similar to that of a plane-parallel capacitor and no components should be found in the x-direction. Near the transition region between the quartz window and the metallic wall the wall currents flow into the waveguide material itself. Measurements of the electric field distribution in an Argon plasma showed deviations from the theoretical field distribution in the plasma [82]. This can be attributed to non-ideal tuning and the orientation of the coupling elements. The field strength into the plasma decays exponentially. Based on the skin effect the penetration depth of the microwaves depends on the permittivity of the plasma and is calculated to be 19 mm using a spatial constant electron density of 10^{11} cm^{-3}, an electron collision frequency of 10^{10} s^{-1} and excitation frequency of 2.45 GHz [81].

The degree of ionisation in such microwave plasmas is of the order of 10^{-5} to 10^{-4}. The electrons are responsible for energy transfer from the external electric field to the plasma particles due to their fast movement because of the collisions with the background gas. Inelastic collisions due to the electron heating process occur at characteristic energy opening up a very efficient energy transfer channel from the electric field to the heavy particles. The plasma is dominated by collisions firstly due to electron-neutral collisions and later collisions between neutrals become more and more important.

Plasmas in the medium pressure range of 10^{-3} to 10 mbar can be generated in the very microwave plasma source in this work. The power density can be varied between 2 and 20 W cm^{-2}. Typical values of $E/_N$ are in the order of 10^{16} V cm^{-2}. In this plasma source electron number densities of 10^{12} cm^{-3} are possible, but practical values of 10^9 to 10^{11} cm^{-3} are found in refs. [83,84].

The electron temperature in the plasma is assumed to be about 1 eV and some tenth higher close to the microwave entrance window before decaying into the vessel [85]. The transport properties of the plasma source are dominated by diffusion processes.

4 Tunable Diode Laser Absorption Spectroscopy (TDLAS)

This method was originally developed for high sensitivity gas monitoring in environmental investigations. Many Molecules have been measured including important atmospheric molecules responsible for the global green house effect and acid rain. More than twenty years ago laboratory experiments were designed to simulate single reactions, such as those in the main reaction paths involved in destruction of atmospheric ozone by chlorine compounds [86]. Several years later, experiments for field use were designed and complete spectrometers were installed in weather balloons or in rockets for measurements in the atmosphere [87,88]. Most experimental set-ups include a long-path multireflection cell for improving sensitivity.

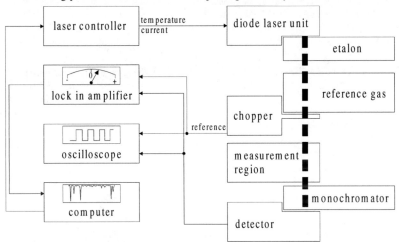

Figure 4.6   Basic tunable diode laser set-up (thick dashed line indicates the path of the laser beam)

The usual set-up of a commercially available TDL spectrometer can be seen in Figure 4.6. The main parts are the diode laser unit, its controller and the detector. The necessary cooling system, either a liquid nitrogen evaporator (LN_2) or a closed cycle helium refrigerator (He), is included in the diode laser unit. The LN_2-system works at temperatures between 80 and 120 K and the He-system provides temperatures between 20 and 120 K. The detectors are usually a HgCdTe semiconductor chip with a matching preamplifier. The chip is normally cooled by liquid nitrogen. Modern arrangements also have closed cooling systems for detectors. Usually the measurement signal is recorded using an oscilloscope or a computer. A lock-in amplifier is not essential for identifing molecular species but is usual for concentration measurements close to the detection limit.

The sensitivity of the diode laser method is based on the use of favourably strong absorption bands in the wavelength region between 3 and 20 μm, corresponding to the wavenumber region of 3300 to 500 cm-1. These absorption bands arise from the characteristic rotational-vibrational fundamental bands of the target molecules. In the near-infrared region between 0.8 and 3 μm overtone or combination bands are often used having, typically, one to several orders of magnitude weaker intensities than the fundamental bands [89]. The mid-infrared region is often called the „fingerprint" region, because of the characteristic spectra of most compounds with up to five atoms and some larger molecules. The high spectral resolution of this method opens the possibility of working also in spectral regions overlapped by atmospheric molecules like water and carbon dioxide because the different target molecules can be separated from the background. Even special absorption bands, so-called „hot bands", due to vibrational excited molecules can be detected.

The spectral linewidth of the laser radiation is very narrow, typically on the order of 10^{-4} cm^{-1} determined by refrigeration vibration and fluctuations in diode temperature and injection current [90]. This linewidth is much smaller than the gas phase Doppler broadening linewidth, which is about 10^{-3} cm^{-1} for diatomic molecules at room temperature.

With respect to the line of sight character of TDLAS the concentration measurement is always restricted to the destination volume formed by the absorption length and the parallel beam diameter which usually varies between 5...15 mm. More spatial information is possible by using Abel inversion, in the case of radial geometry, or special tomography methods.

For measuring the absolute species concentration either the line strength must be known or the molecular gas can be put in a reference gas cell of known partial pressure for calibration. The line strengths for transient species are usually unknown, only a few of them have been measured absolutely. Several mainly stable atmospheric molecules or hydrocarbons were investigated now so the necessary line strengths are available from the literature.

After a short overview of past applications of TDLAS in plasma physics it will be shown how to calibrate the wavelength position and how to calculate the absolute molecular concentration. Other applications of this diagnostic method are also connected to the concentration measurements.

4.1 Historical aspects

The first detection of radicals using infrared TDLAS was published by Menzies et al. in 1977 [86]. They analysed ClO inside a special discharge. The investigation of radicals was combined with pure chemical processes like heating and also with different types of plasmas. The detection of free radicals inside plasmas the characterisation of a large number of transient species while at that stage the plasma properties were not at the centre of interest.

Towards the end of the eighties some groups changed their direction of investigation. The capability of TDLAS was tested as a means of the understanding of plasmas similar to that in industrial application. In particular reaction kinetics was investigated by TDLAS. First experiments were reported in 1987 by Richards et al. [91]. They investigated an rf etching plasma containing chlorine compounds with typical excitation frequencies between 10 kHz and 14 MHz. They measured atomic chlorine by using TDLAS following earlier spectroscopic experiments of Davies and Russell [92]. Other types of etching plasmas were also investigated by TDLAS. The most prominent etching plasmas contain C-F compounds. Therefore Wormhoudt used a typical industrial rf plasma [38]. The results were confirmed and extended in similar plasmas by the group of Kroesen [40,43,93] and also extended in the group of Goto [94,95]. Different source gases were used and all of them could be detected by TDLAS including also different C-F radicals. More information about measured neutral radicals can be found in the review article of Bernath and the book of Hirota [96,97].

Another applied field for TDLAS has been the hydrocarbon-containing plasmas. Ten years ago Wormhoudt as well as Davies and Martineau independently applied this new diagnostic method for carbon deposition reactors using rf plasmas and 20 kHz plasmas, respectively [38,39]. The suitebility for small hydrocarbons was demonstrated. In the following years several groups published their results. Systematic TDLAS measurements of several different hydrocarbons were continued in the group of Davies [41,66,98,99,100]. Goto and co-workers published numerous studies of methyl and methanol concentrations in rf and ECR plasmas under different conditions e.g. investigating the influence of rare gases on the plasma. They have also combined IR absorption with emission spectroscopy, and investigated the effect of water vapour on the methyl radical concentration in argon/methane and argon/methanol rf plasmas using TDLAS [42,67-72]. Kim et al measured CH_3, C_2H_2 and CH_3OH concentrations in methanol/water RF discharges by TDLAS and found that methanol was almost completely dissociated even at medium applied power [73].

Davies and Martineau proposed the extension of TDLAS diagnostics to silane plasmas [41], again representing a new group of molecules.

Beside measurements of the neutral ground state molecules ions can also be detected by a special modulation technique of TDLAS. The velocity modulation technique for the detection of molecular ions was first introduced by Saykally and co-workers [101]. This method specifically allows the detection of absorption lines of charged species in a plasma using the drift velocity of the ions. A wide variety of low molecular weight molecular ions have been detected by TDLAS in purely spectroscopic studies e.g. Si_2^- [102] and SiH_3^+ [103] in silane plasmas. Liu has given a wide survey of molecular ion measurements [104]. Most of these spectroscopic results have yet to be applied in plasma diagnostic studies.

The list of the detected neutral radicals most relevant for $CH_4(CH_3OH)-H_2-Ar-O_2$ microwave plasmas is given in Table 4.1. A more extensive and general list of neutral radicals can be found in refs. [96,97].

mols.	Band	band origin [cm^{-1}]	position range (one useful line position) [cm^{-1}]	references
C$_3$	ν_3	2040	2010 - 2090 (2055.9547)	105,106
C$_5$	ν_3	2169	2155 - 2179	107,108, 109,110
CH$_2$	ν_2		892.61 - 892.79 (892.6421)	111
	ν_2		800 - 911 (856) 1030 - 1173 (1132.0416)	112
CH$_3$	ν_2	606	606 - 750 (606.12)	51
	ν_3		3148 - 3224	52,55
	ν_2	606	(608.3[*1])	65
C$_2$H	ν_3	1840	1789 - 1897 (1849.0907)	113,114
	$\nu_2+\nu_3$ [a]	2091	2057 - 2130 (2089.8)	115
C$_2$H$_3$	ν_7	895	823 - 977 (846.9)	116,117
C$_2$H$_5$			520-550	118
C$_3$H$_2$	ν_3	1277	1250 - 1305 (1258.3)	119
HCO	ν_1	2434	2263 - 2626 (2419.08)	120
	ν_2	1080	1067 - 1100	121,122
	ν_3	1868	1867- 1876 1839 - 1920 (1867.2062)	123,124 125
C$_2$O	ν_1	1870	1964 - 1989 (1984.97)	126
HO$_2$	ν_2	1391	1371 - 1415 (1396.9) (1371.927[*2], 1411.180[*2])	127 128
	ν_3	1097	1081 - 1109 1045 - 1117 (1081.601[*3])	129,130 131,132
OH	rot [b]		508 - 814 (710.88)	133
	rot [b]		918 - 1094 (1072.3807)	134

Table 4.1   *Infrared bands of free neutral radicals accessible to TDLAS in $CH_4(CH_3OH)$-H_2-Ar-O_2 microwave plasmas (a: combination band; b: pure rotational lines; *with line strength [cm molecule^{-1}] 1: 3.2 x 10^{-19} , 2: 1.2 x 10^{-20} , 3: 2.9 x 10^{-21})*

The wavenumber given in brackets show a useful line position for measurements under plasma conditions. Known and promising line strengths are also marked. Most of the radical spectra mentioned in the table have yet to be applied in plasma physics.

4.2   Diode lasers

Laser radiation is emitted in a semiconductor diode due to the population inversion created in the pn-junction as soon as a sufficient current flows. The spectral range of these lasers covers radiation from the visible (0.4 μm) to the infrared (25 μm) depending on the semiconductor material used.

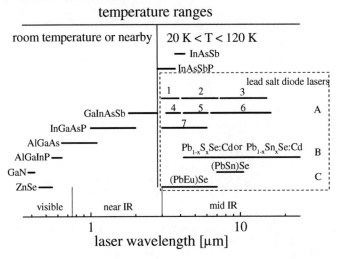

Figure 4.7   *Diode laser material compositions and their corresponding wavelengths [89,135,136]; A- frequency ranges provided by various lead salt diode laser materials (1-$Pb_{1-x}Cd_xS$, 2-$PbS_{1-x}Se_x$, 3-$Pb_{1-x}Sn_xSe$, 4-$Pb_{1-x}Ge_xS$, 5-$Pb_{1-x}Ge_xTe$, 6-$Pb_{1-x}Sn_xTe$, 7-$Pb_xEu_{1-x}Se_{1-y}Te_y$), B- homostructure, C- heterostructure.*

A survey of the materials and their achievable wavelength region can be seen in Figure 4.7. The main distinction between the diodes is due to the type of cooling required. While diodes for the visible and the near-infrared region emit their light at room temperature or nearby, lead salt diode lasers for the mid IR need temperatures between 20 and 120 K. All measurements relevant to this thesis have been performed using mid infrared lead salt diode lasers.

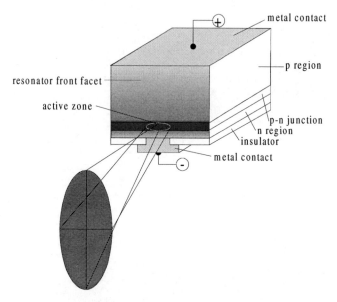

Figure 4.8   Homostructure lead salt diode laser

The basic structure of the diode laser is given in Figure 4.8. The simplest form, the so-called homostructure diode laser, consists of a crystal of PbSnSe or PbSSe semiconductor material with Cd as a donor on which a pn-junction is formed by diffusion. The diffusion zone is characterised by the time of heating during the production process. A rough wavelength determination can be achieved from the salt stoichiometry used and the width of the diffusion zone. With the electrodes placed on the top and bottom a current of several 100 mA flows through the pn-junction initiating the lasing action. In order to convert the stimulated emission into a laser signal an optical cavity is needed. Usually the two parallel resonator facets are polished creating a Fabry-Pérot cavity. The diode laser emission has a characteristic spatial distribution, Figure 4.8, deviating from the Gaussian profile of a HeNe-Laser. The typical thickness of the active zone is about 10 µm and the exiting laser beam is very divergent. The intensity distribution of this radiation is more divergent in the plane perpendicular to the junction plane than that in the plane parallel to the junction plane. This characteristic behaviour requires specific optical elements in front of the diode laser to collimate or focus the infrared beam.

The tuning range of a diode laser can be several hundred wavenumbers. Temperature and current are used for tuning the wavenumber but usually the temperature is set to a certain constant value and the diode is tuned via its current. Direct current can be applied as well as different modulation types. Modulation frequencies between several kHz and GHz can be used

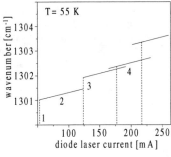

Figure 4.9   Mode structure of a diode laser at a fixed temperature (1-threshold, 2-single mode, 3-mode hop, 4-overlapping modes)

The spectral emission behaviour at different diode currents can be seen in Figure 4.9. The injection current is shown versus the emission wavenumber at a constant temperature. A typical mode structure would be observed as indicated by the dotted lines at corresponding current and wavenumber positions. The tuning range of a single mode is usually about 0.5 to 1 cm^{-1}. Several characteristic properties are shown in this figure. The step like beginning of the emission at the „threshold current" (1) and single mode behaviour (2), where for every set of parameters only one wavenumber is emitted. Further step like changes of the emission wavenumber of about 0.2 to 5 cm^{-1}, so-called „mode hops" (3) and multimode behaviour (4) are also shown. In the latter case the diode laser emits radiation with two or more wavenumbers for one set of parameters. Such multimodes can be separated by a monochromator. The mode structure of diode lasers can change with operation time. The older the diode laser the higher is the threshold current and the mode structure can be shifted to higher wavenumbers or changed completely. This effect can be suppressed so that modern lead salt diode lasers are time stable over weeks [137].

The properties of diode lasers have been further improved resulting in new developments of diode lasers like double heterostructure or buried structure lasers. These diode lasers have a much greater optical gain for given current densities and a significant reduction in threshold currents. Such structures are the most commonly used today.

4.3   Wavenumber calibration

All species are identified by their unique pattern of absorption lines. The spectra of about 42 molecules were determined and their spectral positions and spectroscopic properties are entered in various texts [138] or regularly updated databases HITRAN and GEISA [139,140,141].

Stable and commercially available gases in special glass cells are used as reference sources. The identification of absorption lines of the molecule of interest can either be carried out directly by monitoring the molecule of interest (as a reference gas) or with the aid of one or more other reference gases which absorb radiation at the same or very close wavenumber positions. The latter protocol is used for the identification of radical lines.

The monochromator is needed for filtering out a specific mode and also for the first approximate determination of the wavenumber position. By calibrating the monochromator using the orders of a HeNe laser the diode laser wavenumber can be determined with an uncertainty of 1 cm^{-1}. This is not accurate enough however for the exact wavenumber determination at full spectral resolution. For high resolution measurements an etalon is used together with reference gases. The Fabry-Pérot etalon for infrared radiation consists of a solid Germanium piece with a known free spectral range (FSR).

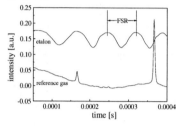

Figure 4.10   *Calibration of spectral positions using a reference gas spectrum and etalon fringes, FSR = 0.0479 cm^{-1} (Free Spectral Range).*

An example of a reference gas spectrum and the etalon signal is shown in Figure 4.10. The identification of lines and the measurement of their absolute positions was carried out using well documented reference gas spectra [138,139,140] and the etalon for interpolation.

Using the very highest purity reference gases avoids problems of identification. The reference gas cells always include a latent water concentration except for gases reacting with water. Atmospheric molecules such as water are recognised in the spectra by their high pressure linewidth caused by the pressure broadening of the absorption line and the length of the absorption path in air. Additional water concentrations in the reference gas cells show a small low pressure peak on top of the high pressure absorption. This behaviour can also be used for the determination of the wavenumber position.

4.4   Concentration measurements

Depending on the actual molecular concentration inside the plasma two different detection techniques were applied: direct absorption and second derivative spectroscopy.

4.4.1   Direct absorption spectroscopy

This technique was used for signals of higher than 90 % transmission. The diode laser signal is modulated by the chopper and demodulated by the lock-in amplifier. By tuning the diode current in very fine steps the spectrum can be recorded on a computer. (see for example Figure 4.6)

The method of direct spectroscopy uses the Lambert-Beer law. (see equation (2.6)) The line shape function which describes the line broadening mechanisms is normalised so that

$$\int\limits_{-\infty}^{+\infty} f(v-v_0)dv = 1 \qquad (4.16).$$

First of all we have natural broadening which is described by a Lorentz profile. Its width is connected with the life time of excited states and is only about 10^{-7} cm^{-1}. This value is far lower than that of the other broadening widths and can be safely neglected. So, Doppler and pressure broadening are the dominant processes. Doppler broadening can be described by a Gaussian profile and dominates the line shape at pressures below 10 Torr. This line shape function is,

$$f_D(v-v_0) = \frac{1}{\Delta v_D}\sqrt{\frac{\ln 2}{\pi}}\exp\left[-\ln 2\left(\frac{v-v_0}{\Delta v_D}\right)^2\right] \qquad (4.17)$$

where Δv_D is the half width at half maximum height (HWHM) of the line. This is given by

$$\Delta v_D[cm^{-1}] = \frac{1}{c}\sqrt{2\ln 2kN_A}\ v_0\sqrt{\frac{T}{M}} = 3581\,10^{-7}v_0[cm^{-1}]\sqrt{\frac{T[K]}{M[g\,mol^{-1}]}} \qquad (4.18)$$

where k is the Boltzmann constant, N_A is the Avogadro's number, T is the temperature and M is the molecular weight. For practical tasks the second equation is very useful. The pressure broadening due to collisions of the molecules can be described by a Lorentzian profile. It becomes significant above 10 Torr. This line shape function is given by,

$$f_L(v-v_0) = \frac{1}{\pi}\frac{\Delta v_L}{(v-v_0)^2 + (\Delta v_L)^2} \qquad (4.19)$$

where Δv_L is the half width at half maximum height (HWHM) of the line. Its value is approximately

$$\Delta v_L = \Delta v_{L_0}\frac{p}{p_0}\sqrt{\frac{T_0}{T}} \qquad (4.20)$$

where p and T are the partial pressure and the temperature respectively of the gas, p_0 and T_0 the standard pressure and temperature and Δv_{L0} is the Lorentz width at standard conditions [142]. Generally, the actual line shape of a transition contains both the Gaussian and Lorentzian shapes. The resulting folded line shape function can be described by a Voigt profile.

All experiments were done at low pressure (p<10 Torr) so that pressure broadening can be neglected. Using the Lambert-Beer law the species concentration n can than be calculated as follows

$$n[cm^{-3}] = 2.13\ \ln\left(\frac{I_0}{I}\right)\frac{\Delta v_D[cm^{-1}]}{S[cm]\,l[cm]} \qquad (4.21)$$

whereas I_0 is the diode laser intensity entering the plasma, I is the transmitted intensity, S the line strength and l the absorption path length.

4.4.2   Second derivative spectroscopy

This technique provides better sensitivity by up to two orders of magnitude because of the significant improvement of the signal to noise ratio (S/N) using phase sensitive detection of the lock-in amplifier. The second derivative technique was applied when the transmission was 10 % or lower. The modulation of the injection current has been widely used to measure concentrations. The modulation frequency has been varied from the MHz region, where the technique is called FM spectroscopy [143], to kHz what was used for the second derivative spectroscopy applied for several of the measurements of this thesis.

The experimental arrangement and the components are shown in Figure 4.6. The diode laser current itself is modulated with a small amplitude signal. At the same time the direct current of the diode laser is tuned in μA-steps to record the spectrum.

The absorption line profile can be described by the line shape function given earlier and the modulated detector signal can be expressed as

$$U_{ds} = f\left(v + U_m - v_0\right) \tag{4.22}$$

where U_m is the modulation signal. In case of sinusoidal wave modulation the detected signal is given by

$$U_{ds} = f\left(v + A_m \sin(\omega t + \varphi) - v_0\right) \tag{4.23}$$

where ω is the frequency and A_m is the amplitude of the modulation. The modulation amplitude should be small compared to the halfwidth of the line profile. More detailed descriptions of the parameters used for derivative spectroscopy can be found in ref. [144].

The signal can be expanded as a Taylor series:

$$U_{ds} = f\left(v - v_0\right) + A_m \sin(\omega t + \varphi) \frac{\partial f\left(v - v_0\right)}{\partial v} \tag{4.24}.$$
$$+ \frac{1}{2}\left(A_m \sin(\omega t + \varphi)\right)^2 \frac{\partial^2 f\left(v - v_0\right)}{\partial v^2} + \dots$$

If the terms in $\sin(n\omega t + \varphi_n)$ can be grouped together then the first and second derivatives are proportional to the first and second harmonics detected by the lock-in amplifier. In the case of a sinusoidal reference signal the $1f$ and $2f$ signals are proportional to the first and the second derivatives of the line profile [104]. Experiments described in this thesis has been carried out using only second derivative spectroscopy. Higher order derivatives can also be used, e.g. to increase the wavelength resolution [145].

Although quantitative measurements are desirable the derivative technique only provides relative values of the concentration. Therefore the signal amplitude must be calibrated by a direct measurement of the molecule concentration whether inside a reference gas cell at a known partial pressure or in the plasma under the same conditions. Such measurements were reported for the methyl radical using N_2O gas in ref. [98]. The S/N ratio was about 30:1 for the best conditions. If fringes caused by back reflections are of the order of the signal amplitude the recording of a baseline and its subtraction from the measurement signal helps to correct the signals and improve the S/N ratio further.

5   Experimental Arrangement

The general arrangement of the plasma source and the tunable diode laser systems used for all experiments are shown in Figure 5.11.

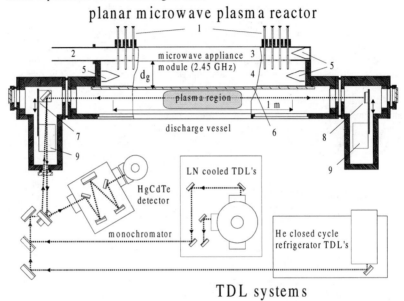

Figure 5.11   *Experimental arrangement of planar microwave plasma reactor (side view) and tunable diode laser (TDL) infrared sources. The path of the diode laser beam is indicated by dotted lines. 1- tunable coupling elements, 2- to generator, 3- upper waveguide, 4- interface waveguide, 5- absorber, 6- microwave window, 7- periscope optics, 8- retroreflector, 9- stepper motor.*

Two additional vacuum side ports are mounted at both front ends of the discharge vessel. These ports contain the periscope optics (7) and the retroreflector (8) moved by stepper motors (9) for vertical movement of the optical axis (7) [142]. Stable molecules are assumed to be distributed homogeneously in the optical path and transient molecules like the methyl radical are assumed to exist only in the plasma region. Two different diode laser assemblies were applied using the same optical path through the plasma source. Both the monochromator and the detector were used for the two diode laser systems.

5.1 Reactor system

The main features of the concept of planar microwave discharges was described in chapter 3.2. Only the specific properties of the planar microwave plasma used here are given [82]. The design of the plasma source is shown in Figure 5.11 and the numbers in brackets refer to this diagram. The applicator contains two T-shaped rectangular waveguides (3,4) to guide the microwaves into the discharge chamber. The upper waveguide (3) represents a travelling-wave slotted waveguide for distributing the microwave power over the length of the microwave window (6). Coupling holes in the contact surface between the slotted and the interface waveguide (4), with a distance of $\lambda/4$ are used together with tunable coupling elements (1) inside the holes for improving the tuning range and the efficiency of the microwave power coupling behaviour. The metal walls alongside the gap d_g, the slotted surface and the microwave window (6) with the conducting plasma below form the interface waveguide (4). Microwaves can propagate in both directions to the ends of the waveguides. Therefore specific absorbers (5) are installed for damping reflections. The dimensions of the discharge vessel used here were $120 \times 21 \times 15$ cm. In principle, the power transfer surface to the plasma chamber can be extended to a rectangular surface of about 600 cm^2, but only about one sixth of it was used for the experiments described in this thesis. Microwave power up to 6 kW can be applied, but 1.5 kW was used only. This plasma source can be used over a medium range of pressure from 10^{-2} to 10 mbar. The measurements were performed at a microwave frequency of 2.45 GHz and power of 1.5 kW. The reflected power was shown to be about 10 %. In addition power is lost due to heating of the vessel walls and absorbers. The effective applied power yielded 1.2 kW. This corresponds to a power flux in the TDL observation plane of about 12 W cm^{-2}. This planar discharge configuration has the advantage of being well-suited for end-on spectroscopic observations because considerable homogeneity can be achieved over relatively long plasma path lengths [146].

Gas flows were arranged by mass flow controllers and the gases were mixed before entering the reactor. They were pumped out via a port in the reactor wall. For all gases the highest available purity was used to avoid interferences from unknown species. The gas mixture supplying the reactor consisted of x sccm H$_2$+ 60 sccm Ar + y sccm O$_2$ + z sccm CH$_4$ or CH$_3$OH. The total flow rate and pressure were kept constant at 555 sccm and 1.5 mbar respectively. The pumping speed was adjusted with a butterfly valve to maintain a constant pressure in the plasma and the pressure measured on a capacitance pressure gauge. This was the situation for flowing conditions (fc). For both methane and methanol containing mixtures four different flow rates (z) were selected, 5, 15, 25 and 40 sccm corresponding to 0.9, 2.7, 4.5 and 7.2 % of the total flow rate respectively. When oxygen was added the proportion of hydrogen was reduced correspondingly to maintain a constant flow rate.

In order to eliminate the influence of hydrodynamic processes inside the plasma vessel and side ports an empirically determined method was developed for the concentration measurement of stable species under static conditions (sc). The plasma reactor was first thoroughly evacuated and then filled with the flowing gas mixture of interest. Once the pressure had been stabilised the microwave generator was switched on and after 40 seconds the concentration measurements were made under flowing conditions. With the plasma on, the gas flow and the vacuum pump were stopped simultaneously to isolate the reactor while maintaining constant pressure as far as possible (variations were < 15%). After another minute concentration measurements in the closed reactor were made. Lastly the measurement was repeated with the plasma off to study the influence of temperature changes on the line strength of the absorption lines.

5.2 TDL assemblies

Two different TDL spectrometers have been used, together with a single monochromator placed in front of the detector. The main ideas behind using these semiconductor radiation sources were explained in chapter 4.2. The one special feature of the set-up was the use of a single optical path for both systems so changing the systems without loosing much time. The main features of the TDL systems used are given in Table 5.2.

Properties	He closed cycle refrigerator	LN$_2$ evaporator
supplier	Laser Photonics (Aero Laser)	IPM Freiburg
working temperature	20 - 100 K	80 - 120 K
diode laser current	up to 1000 mA	up to 2000 mA
modulation types	triangular, rectangular, sinusoidal, ramps	triangular, rectangular, sinusoidal, ramps
modulation frequency range	up to 300 kHz	up to 5 MHz
number of installed diodes	4 (in one vessel)	2 (in single vessels)
accessible wavenumber area	3300 - 500 cm^{-1}	3300 - 1000 cm^{-1}
diode change time	1.5 days	1.5 hours

Table 5.2   Main features of the He closed cycle refrigerator and the LN$_2$ evaporation system

The diode lasers are mounted in vacuum vessels and cooled down to temperatures between 20 K and 120 K. The cooling principle used in both systems is different. One is a closed cycle Helium refrigerator which can reach working temperatures of about 20 K. The other cooling system uses the evaporation of liquid nitrogen (LN$_2$) and the minimum working temperature is therefore about 80 K. While the reservoir of LN$_2$ must be refilled daily the closed cycle Helium refrigerator is able to work around the clock for weeks.

Both systems were aligned on the same optical path. The infrared diode laser beam from one of two TDL source assemblies (Figure 5.11) entered the plasma chamber via a KBr window. Its distance from the microwave entry window was adjusted using a periscope mirror system in combination with a retroreflector. The great majority of measurements reported here were made at a fixed distance of 20 mm from the microwave entry window with the laser beam making two passes through the plasma. After leaving the plasma reactor the diode laser beam first passed through a mode selection monochromator before detection with a HgCdTe detector. Two different gratings were used with blaze wavelengths of 6 μm and 15 μm. Two detectors with maximum wavenumber sensitivities at 10 μm and 16 μm were used. To avoid the influence of thermal energy the detector chip itself is cooled by liquid nitrogen.

6   Experimental Results

The plasma chemistry of molecular microwave discharges using source gas mixtures of H_2-O_2-Ar-CH_4 or -CH_3OH was investigated by infrared TDLAS. The methyl radical and some related hydrocarbons were assumed to be produced based on former investigations, e.g. refs. [62,66]. In addition some oxygenated compounds were also expected to be present.

The calibration of the wavenumber position was described in chapter 4.3 and the calculation of the species concentration can be found in chapter 4.4.

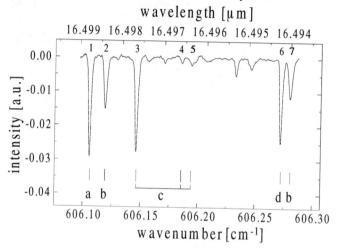

Figure 6.12   TDL absorption spectrum of some molecules in a H_2-O_2-Ar plasma containing methanol under flowing conditions. Legend: a- CH_3OH, b- CH_3, c- N_2O, d- CO_2. The line due to N_2O is from a calibration cell placed in the beam path.

line number	molecule	line	line position [cm⁻¹]	line strength [cm]	references
1	CH_3OH		606.082		
2	CH_3	Q(3,3)	606.12032	5.8×10^{-19}	[51,65]
3	N_2O	R(19)	606.146055	6.05×10^{-22}	[138,139]
4	N_2O	R(19)	606.188708	5.6×10^{-23}	[138,139]
5	N_2O	R(19)	606.197530	5.6×10^{-23}	[138,139]
6	CO_2	P(15)	606.276994	2.722×10^{-21}	[138,139]
7	CH_3	Q(4,4)	606.28494		[51]

Table 6.3   The characteristics of identified lines corresponding to Figure 6.12

A representative TDL absorption spectrum of a microwave plasma (H_2-O_2-Ar-CH_3OH, p = 1.5 mbar, P = 1.5 kW) with lines due to methanol, methyl and carbon dioxide in the plasma is shown in Figure 6.12. The lines of N_2O come from a reference gas cell placed in the beam path.

The present TDLAS study achieved a systematic, quantitative investigation of eleven molecules in hydrocarbon-containing microwave plasmas. Beside the methyl radical the stable molecules CH_4, CH_3OH, C_2H_2, C_2H_4, C_2H_6, CH_2O, HCOOH, CO, CO_2 and H_2O were detected in mixed H_2-Ar-O_2 plasmas containing small percentages (0.9 to 7.2%) of methane or methanol.

molecule line		line position [cm^{-1}]	line strength [cm]	refs.	remarks
CH_3	Q(3,3) Q(4,4)	606.12032 606.28494	5.8 x 10^{-19}	[51,65]	calculated from Q(8,8)
CH_4		1301.5539 1301.66448 1302.78744	3.87 x 10^{-20} 4.08 x 10^{-21} 4.50 x 10^{-20}	[139] [139] [139]	CH_4 plasmas CH_4 plasmas CH_3OH plasmas
CH_3OH		1302.667	9.1 x 10^{-22}		derived from CH_3OH concentration in the plasma source
C_2H_2	P(15) P(11)	1293.5213 1302.5968	8.43 x 10^{-20} 1.154 x 10^{-19}	[139] [139]	CH_4 plasmas CH_3OH plasmas
C_2H_4		948.436157	3.53 x 10^{-20}	[140]	
C_2H_6		806.404785	3.54 x 10^{-22}	[140]	
CO	P(22)	2050.854	1.804 x 10^{-20}	[140]	
CO_2	P(15)	606.276994	2.722 x 10^{-21}	[139,140]	
H_2O		1305.48859	1-4 x 10^{-22}	[139]	strongly dependent on T
CH_2O	R(13)	1774.7961	5.38 x 10^{-20}	[147,148]	
HCOOH		1117.22583	2.01 x 10^{-20}	[139]	
N_2O	R(19)	606.146055	6.05 x 10^{-22}	[139]	reference gas

Table 6.4   Absorption lines used for molecular identification and concentration measurements.

Table 6.4 gives the absorption lines used for concentration measurements, their positions and line strengths. The plasma length varied between 11 and 18 cm depending on the discharge conditions. For the purpose of concentration measurements the methyl radical was assumed to exist only in the plasma region of the reactor. The line strength of the Q(3,3) transition of the methyl radical was calculated from the Q(8,8) line, measured by Wormhoudt [65] and adjusted for the higher temperature in the active plasma zone. On the other hand stable molecules were assumed to diffuse into the side ports, through which the diode laser beam passed. Therefore, for their concentration measurements the whole optical path of 450 cm inside the reactor was used. The temperature of the plasma was determined by OES [149,150,151,152]. A value of 1000 ± 100 K in the case of a high hydrogen content was found. Outside the plasma region a temperature of 300 K was used for the calculation of the concentration of the stable species. The lines used to deduce the concentrations of stable molecules were turned out to be relatively insensitive to temperature under the conditions used. An exception was the H_2O absorption line. In this case the temperature dependence of the line strength was estimated empirically.

6.1 General observations on species concentrations at various oxygen contents

The concentrations of eleven species measured at O_2 flows of up to 455 sccm by TDLAS are shown for flowing conditions in Figure 6.13 (for methane) and Figure 6.14 (for methanol). The corresponding results for static conditions are shown in Figure 6.15 and Figure 6.16. The data for the product C-2 hydrocarbons are shown in the right hand panels of each figure. Dotted lines have been included as a guide to the eye. The data were acquired at constant microwave power and total pressure of 1.5 kW and 1.5 mbar, respectively. The flow used in Figure 6.13 - Figure 6.16 for methane and methanol (40 sccm = 7.2 %) was selected to represent typical flows used in practical CVD reactors.

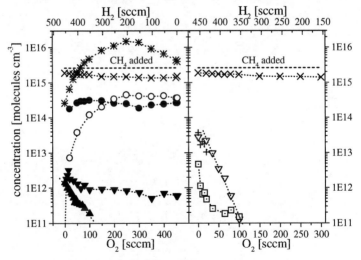

Figure 6.13   *Molecular concentrations in a methane (fc) plasma as a function of oxygen flow rate measured by TDLAS. Data points have been joined by dotted lines as a guide to the eye. For clarity the C-2 product hydrocarbons are shown separately in the right hand panel. Legend:* ✕ - CH_4; ▽ - C_2H_2; □ - C_2H_4; + - C_2H_6; ● - CO; ○ - CO_2; ▲ - CH_3; ✳ - H_2O; ▼ - CH_2O. [44]

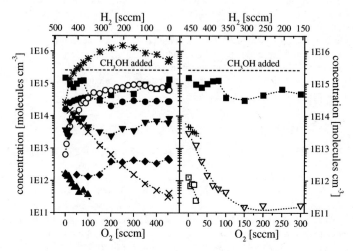

Figure 6.14 *Molecular concentration as a function of oxygen flow rate in a methanol (fc) plasma measured by TDLAS. Data points have been joined by dotted lines as a guide to the eye. For clarity the C-2 product hydrocarbons are shown separately in the right hand panel. Legend: ✕ - CH_4; ∇ - C_2H_2; □ - C_2H_4; + - C_2H_6; ● - CO; ○ - CO_2; ▲ - CH_3; ✳ - H_2O; ▼ - CH_2O; ■ - CH_3OH; ◆ - HCOOH. [44]*

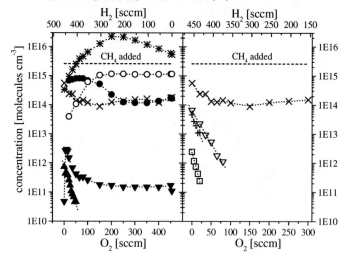

Figure 6.15 *Molecular concentration as a function of oxygen flow rate in a methane (sc) plasma measured by TDLAS. Legend: ✕ - CH_4; ∇ - C_2H_2; □ - C_2H_4; + - C_2H_6; ● - CO; ○ - CO_2; ▲ - CH_3; ✳ - H_2O; ▼ - CH_2O. [44]*

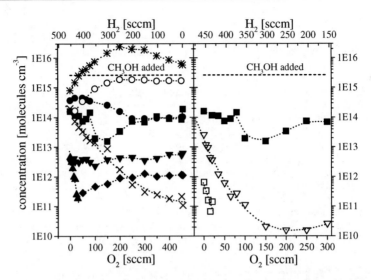

Figure 6.16   Molecular concentration as a function of oxygen flow rate in a methanol (sc) plasma measured by TDLAS. Legend: ✕ - CH_4; ▽ - C_2H_2; ☐ - C_2H_4; + - C_2H_6; ● - CO; ○ - CO_2; ▲ - CH_3; ✳ - H_2O; ▼ - CH_2O; ■ - CH_3OH; ◆ - HCOOH. [44]

Figure 6.13 to Figure 6.16 give an overview of the mass balance and degree of dissociation, as well as product concentrations ranging over six orders of magnitude. Some useful generalisations can be made from this data. The reaction products are more uniformly distributed over the concentration range for methanol than for methane plasmas. There are some differences between the species concentration profiles measured in flowing and static conditions. The same products are not always found for both precursor hydrocarbons. Thus methane was detected in methanol plasmas but methanol was not detected in methane plasmas. Lastly, as shown in the right hand panels of each figure, the most noticeable oxygen dependent changes occurred for the minor constituents at low oxygen concentrations. The wide range of concentrations measured for the different species provides a convenient means for detailed discussions below. The highest concentrations are for the precursor molecules and the major product H_2O. Next come the two most stable products of oxidation CO and CO_2, and finally the minor constituents of most interest here, namely the stable and unstable hydrocarbons (C_2H_2, C_2H_4, C_2H_6 and CH_3). The other minor oxidation products investigated were CH_2O and HCOOH.

The measurement error has also been determined. The relative uncertainty of all measurements was dependent on the signal to noise ratio inside the plasma and also on the flow conditions. Different absolute molecule concentrations can also arise from variations of the absorption length, the neutral gas temperature and the line strength. For optimum performance TDLAS should use absorption lines with a reasonable line strength compared to the expected order of magnitude of the molecule concentration. The experimental transmission should be 10 % or lower, and saturated lines should be avoided. S/N ratios of about 100/1 or better are ideally required. Measurements of the spectra from reference gas cells yielded statistical errors of less than 1 %. These small fluctuations may be due to mode structure from the diode temperature or injection current instabilities. This value is smaller than the figure of 5 % found for the measurements of the spectra of the plasma. In this case we can expect homogeneity problems of spatial distribution of the molecules, especially when changing conditions from fc to sc. The microwave field structure can change, as noted by changes in the visible regions inside the discharge followed by corresponding changes in the molecular concentrations. Related to this behaviour are also changes of the neutral gas temperature of the plasma. If the absorption signals during a series of measurements decrease the statistical uncertainty than increase up to 15 % because the S/N ratio becomes poor. If the absorption signals are low the situation can be rectified by changing the detection technique from the direct absorption to derivative spectroscopy. Filtering processes increase the S/N ratio and therefore decrease the measurement error. The main factors which contribute directly to the uncertainty of the absolute concentration measurement are the absorption length and the line strength. All line strengths used here were taken from the international data bases HITRAN and GEISA [139,140].

6.2 Degree of hydrocarbon dissociation

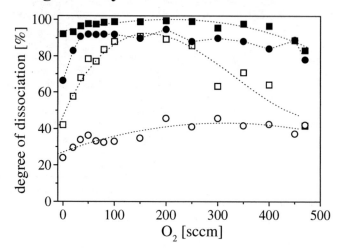

Figure 6.17   *The degree of dissociation (%) of methane (○ , ●) (fc,sc) and methanol (□,■) (fc,sc) as the content of oxygen was increased. The proportion of methane and methanol added was the same (25 sccm = 4.5%) in all cases. [44]*

A knowledge of the degree of dissociation of the hydrocarbon precursors is necessary in order to study the conversion of the added gas mixture into products and for understanding the mass balance of the plasma processes. Figure 6.17 shows the degree of dissociation of methane and methanol at increasing oxygen flow rates under flowing and static conditions (4.5% hydrocarbon ≡ 25 sccm). It can be seen that increasing oxygen flows have a more marked effect for methanol than for methane plasmas. Under static conditions (●,■) the degree of dissociation is higher, as expected, and reaches 99% for methanol and 94% for methane plasmas.

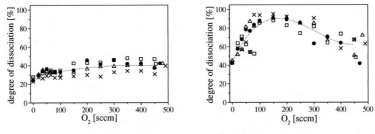

Figure 6.18   *The degree of dissociation for four different hydrocarbon flows (fc) using methane (left panel) and methanol (right panel). The dotted lines represent the mean values of the four different hydrocarbon flows. Flow rates: □ - 40 sccm = 7.2 %;*
● *- 25 sccm = 4.5 %; △ - 15 sccm = 2.7 %, ✕ - 5 sccm = 0.9 %. [153]*

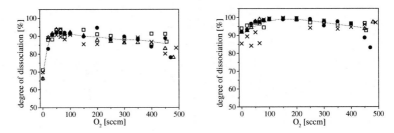

Figure 6.19   *The degree of dissociation for four different hydrocarbon flows (sc) using methane (left panel) and methanol (right panel). The dotted lines represent the mean values of the four different hydrocarbon flows. Flow rates: □ - 40 sccm = 7.2 %;*
● *- 25 sccm = 4.5 %; △ - 15 sccm = 2.7 %, ✕ - 5 sccm = 0.9 %.*

The other hydrocarbon admixtures can be seen in Figure 6.18 and Figure 6.19. At other hydrocarbon concentrations the degree of dissociation showed, within error limits, a similar dependence on the partial pressure of oxygen. The dotted lines in these figures represent the mean values of all four hydrocarbon admixtures.

Although the applied microwave power was kept constant the plasma region inside the reactor contracted as the oxygen concentration was increased. It can be safely assumed that the electron energy distribution in the plasma has changed hence affecting the degree of dissociation of the precursor hydrocarbons. The total dissociation cross section for methane has been reported and for low energy electrons (< 20 eV) the cross section increases approximately linearly with electron energy [154]. Unfortunately methanol dissociation cross sections have not been reported but by analogy with methane an increase in cross section with electron energy can be expected.

6.3 Methyl radical concentration

The CH_3 radical was monitored using the Q(3,3) transition of the υ_2 fundamental band at 606.12 cm^{-1}[51]. The peak height of the second derivative signal was calibrated either against a direct concentration measurement or against the R(19) line of N_2O at 606.146 cm^{-1} which has a known line strength. The concentration of the methyl radical was then calculated using the method described in chapter 4.4.2. It can be seen from Figure 6.13 to Figure 6.16 that the methyl radical has one of the lowest concentrations of any species in the plasma, over three orders of magnitude lower than the hydrocarbon precursors added. Figure 6.20 shows the variation of the methyl radical concentration in the absence of oxygen as a function of added methane or methanol. As expected, flowing conditions give much higher concentrations than static conditions. For 40 sccm of methanol [fc]/[sc] ~ 4. With the exception of methane (fc) the remaining data in Figure 6.20 increase nearly linearly with precursor concentration.

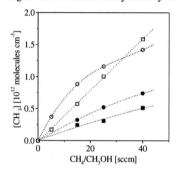

Figure 6.20 *Variation of the concentration of the methyl radical as a function of methane or methanol added to H_2-Ar plasmas measured by TDLAS. The data have been fitted to a second order polynomial. Legend: $\circ$ - CH_4, fc; $\square$ - CH_3OH, fc; $\bullet$ - CH_4, sc; $\blacksquare$ - CH_3OH, sc. [44]*

Similar results have been reported for argon microwave plasmas containing low methane partial pressures (70 sccm Ar + (0 - 40) sccm CH_4; p= 500 Pa) [149]. In addition the behaviour for methane (fc) is similar to the curve obtained by Naito et al. in a rf methane discharge containing a trace of argon [68]. The measurements reported by Röpcke et al. were carried out at lower total flow rates (70 - 110 sccm) than those reported here (555 sccm) [149]. They detected a decrease in the methyl radical concentration in the direction normal to the microwave input window (Figure 5.11). Such a concentration gradient was not detected for the much higher flow rates used in the experiments described in the present study.

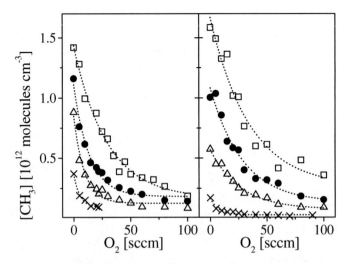

Figure 6.21   *Decay of the methyl radical concentration as a function of added oxygen measured by*
TDLAS. Left panel - methane plasma (fc); right panel - methanol plasma (fc). The
symbols have the same meaning in both panels. Flow rates: □ - 40 sccm; ● - 25 sccm;
△ - 15 sccm, ✕ - 5 sccm. [44]

When oxygen was added to methane or methanol plasmas the methyl concentration was
found to decay exponentially as the oxygen content increased. Figure 6.21 shows the
results for a wide range of fixed hydrocarbon input flows. The dotted lines in Figure
6.21 are the exponential curves from fits to the data points. The methyl radical
concentration can be seen to be slightly higher for methanol than for methane at 40 sccm
with the opposite effect when the hydrocarbon flow is 5 sccm.

6.4   Stable C-2 hydrocarbons and oxygen-containing species

The concentrations of C_2H_2, C_2H_4 and C_2H_6 as well as formaldehyde and formic acid
have already been shown as a function of the oxygen flow rate in the survey figures
Figure 6.13 to Figure 6.16. The following more detailed diagrams of the molecule
concentrations show the results for the different amounts of hydrocarbon and emphasize
the characteristic behaviour.

6.4.1 Acetylene, ethylene and ethane

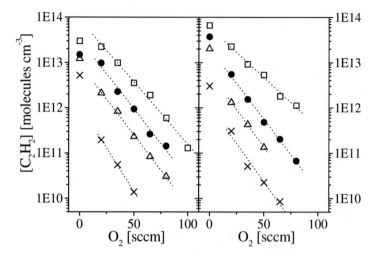

Figure 6.22   *Variation of acetylene concentration with oxygen content using methane-containing source gas (left panel - fc, right panel - sc). Flow rates:* □ *- 40 sccm;* ● *- 25 sccm;* △ *- 15 sccm,* ✗ *- 5 sccm.*

The behaviour of the acetylene concentration can be seen for the methane-containing source gas in Figure 6.22 and for the methanol-containing source gas in Figure 6.23. The left hand and upper panels give the results for fc and the other panels for sc. For increasing oxygen content an exponential concentration decay over two orders of magnitude was observed for source gas mixtures containing methane. The acetylene concentration could be measured up to 100 sccm of oxygen. There are no significant differences between fc and sc, but the acetylene concentration in sc is slightly higher for the 40 and 25 sccm methane flow. This may be a sign that acetylene is a final product of a reaction chain, especially in the case of low oxygen content. The major path for acetylene production in a plasma is via electron impact dissociation of ethylene. In the microwave plasma where the gas temperature can reach 1000 K, hydrogen abstraction and additional reactions with hydrocarbons also contribute to the production of acetylene.

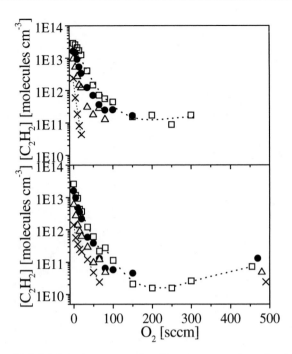

Figure 6.23   *Variation of acetylene concentration with oxygen content using methanol-containing source gas (upper panel - fc, lower panel - sc); Flow rates:* □ *- 40 sccm;* ● *- 25 sccm;* △ *- 15 sccm,* ✗ *- 5 sccm.*

For the methanol-containing source gas the acetylene concentration decays exponentially with oxygen contents of up to 100 sccm in the same way, Figure 6.23. This behaviour could be found for both conditions, fc and sc. The acetylene concentration becomes nearly constant with higher oxygen content, at least for the highest methanol flow rate. Under fc the acetylene concentration fell below the detection limit at oxygen contents of 300 sccm, while under sc a slightly increasing amount of acetylene was found. Finally under sc even without additional hydrogen acetylene is detectable. This fact also supports the idea that the acetylene molecule is a final product even under sc with high oxygen content, but with a very low concentration compared to H_2O, CO_2 and CO (Figure 6.16). The measurement of the acetylene concentration was limited by discharge instabilities, optical fringes or overlapping of other spectral lines as for the methanol-containing source gas. Nevertheless the line strength of the acetylene absorption line used allowed concentration measurements down to 8×10^9 molecules cm^{-3}.

The behaviour of the ethylene concentration is shown in Figure 6.24 for the methane case. The left panel shows the value for fc and the right panel that for sc.

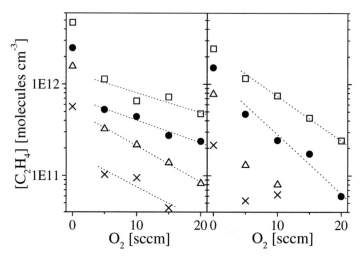

Figure 6.24   Variation of ethylene concentration with different oxygen content in methane-
containing plasmas; left panel - fc, right panel - sc. Flow rates: □ - 40 sccm;
● - 25 sccm; △ - 15 sccm, ✕ - 5 sccm.

When H_2-O_2-Ar source gas mixtures with small amounts of both hydrocarbons, methane
or methanol, were used the measured concentration of ethylene decreased exponentially
with increasing oxygen content. The ethylene concentration for fc is always higher than
that for sc, so that ethylene is converted more to other products, e.g. to acetylene. The
ethylene concentration using methane-containing source gas is always slightly higher
than that using methanol. The highest ethylene concentration was found only when
oxygen was absent showing that reactions with oxygen contribute to the loss of
ethylene. Its concentration for the methanol-containing source gas could only be
measured for the two highest methanol flow rates. At lower methanol content the
ethylene concentration fell below the detection limit of about 5×10^{10} molecules cm^{-3}.

In the case of all measured plasma-produced hydrocarbons the same interesting result
was found for the first data point. Using methanol admixtures the concentration value
without additional oxygen admixtures fell the error limits on the curve for the
exponential decays. A different situation was found for the methane-containing source
gas. The first concentration value using the source gas mixture without oxygen content
was more or less well removed from the extrapolated exponential decay curves. It
showed that the oxygen free discharge is different in this case and this first data point
was excluded for the determination of the exponential decay rates.

The results are also given in tabular form in Table 6.5 and Table 6.6 and include those
for the methyl radical.

Table 6.5 Decays of the product hydrocarbons in methane plasmas. For each % of methane the concentrations are given without added oxygen (n[O$_2$]= 0), the minimum detected concentration on the exponential decay line with the % O$_2$ added (n$_{min}$/ O$_2$ [%]), and the % added O$_2$ required to cause a decay in concentration of one order of magnitude (n decay O$_2$ [%]). n: molecules cm^{-3}; fc: flowing conditions; sc: static conditions. [44]

Species		0.9% n [O$_2$]=0	0.9% n$_{min}$ O$_2$ [%]	0.9% n decay O$_2$ [%]	2.7% n [O$_2$]=0	2.7% n$_{min}$ O$_2$ [%]	2.7% n decay O$_2$ [%]	4.5% n [O$_2$]=0	4.5% n$_{min}$ O$_2$ [%]	4.5% n decay O$_2$ [%]	7.2% n [O$_2$]=0	7.2% n$_{min}$ O$_2$ [%]	7.2% n decay O$_2$ [%]	average n decay O$_2$ [%]
CH$_3$	fc	3.7E11	1.0E11 / 2.7	10.8±1.5	8.8E11	2.1E11 / 4.0	12.25±1.2	1.2E12	3.1E11 / 4.0	13±1.0	1.4E12	4.4E11 / 10.8	14.6±1.3	**12.7**
	sc	-	- / -	-	3.2E11	4.8E10 / 3.6	4.7±0.5	5.2E11	1.1E11 / 3.6	5.8±0.5	7.4E11	4.7E10 / 9	6.9±0.6	5.8
C$_2$H$_2$	fc	5.1E12	1.4E10 / 9.0	4.7±0.1	1.2E13	2.9E+10 / 14.4	5.8±0.1	1.5E13	1.1E11 / 14.4	5.9±0.4	3.0E13	1.5E11 / 18.1	6.5±0.3	5.7
	sc	3.0E12	7.4E09 / 11.7	5.4±0.7	2.2E13	1.3E11 / 9.0	5.4±0.1	3.7E13	6.4E10 / 14.4	5.7±0.2	6.6E13	1.0E12 / 14.4	8.1±0.5	6.1
C$_2$H$_4$	fc	5.7E11	5.0E10 / 2.7	4.9±2.3	1.6E12	8.5E+11 / 3.6	4.5±0.2	2.5E12	2.3E11 / 3.6	7.2±1.1	4.7E12	4.9E11 / 9.0	8.3±2.8	6.2
	sc	2.1E11	- / -	-	7.8E11	- / -	-	1.5E12	6.9E10 / 3.6	3.2±0.5	2.4E12	2.5E11 / 9.0	3.9±0.2	3.6
C$_2$H$_6$	fc	-	- / -	-	-	- / -	-	1.4E13	- / -	-	3.7E13	1.3E13 / 3.6	13.3±10.2	**13.3**
	sc	-	- / -	-	-	- / -	-	-	- / -	-	4.6E13	9.9E12 / 3.6	6.5±1.8	6.5

Table 6.6 *Decay of the product hydrocarbons in methanol plasmas. Complete legend as for Table 6.5. [44]*

Species		% methanol in methanol/oxygen/argon/hydrogen plasmas												
		0.9%			2.7%			4.5%			7.2%			
		n $[O_2]=0$	n_{min} O_2 [%]	n decay O_2 [%]	n $[O_2]=0$	n_{min} O_2 [%]	n decay O_2 [%]	n $[O_2]=0$	n_{min} O_2 [%]	n decay O_2 [%]	n $[O_2]=0$	n_{min} O_2 [%]	n decay O_2 [%]	**average** ***n* decay** O_2 [%]
CH_3	fc	1.7E11	3.5E10 5.4	$15.8_{\pm 5.0}$	5.7E11	2.1E11 5.4	$16.5_{\pm 1.2}$	1.0E12	3.3E11 7.2	$18.0_{\pm 1.1}$	1.6E12	4.2E11 10.8	$20.2_{\pm 1.6}$	**17.6**
	sc	-	- -	-	2.4E11	4.8E10 3.6	$4.3_{\pm 0.6}$	3.1E11	7.1E10 3.6	$4.7_{\pm 0.6}$	5.1E11	1.9E10 5.4	$3.7_{\pm 0.5}$	4.2
CH_4	fc	3.3E13	9.9E11 18.0	$16.8_{\pm 3.2}$	7.5E13	5.8E12 18.0	$17.3_{\pm 1.1}$	1.2E14	1.2E13 18.0	$16.9_{\pm 1.2}$	1.8E14	1.2E13 27.0	$21.3_{\pm 1.0}$	**18.1**
	sc	5.3E13	4.0E12 18.0	$15.1_{\pm 3.6}$	8.9E13	3.3E12 18.0	$17.3_{\pm 1.1}$	1.5E14	4.9E12 18.0	$16.5_{\pm 2.0}$	2.0E14	1.5E13 14.4	$19.3_{\pm 2.0}$	**17.1**
C_2H_2	fc	2.4E12	3.9E10 3.6	$2.4_{\pm 0.2}$	9.5E12	2.9E11 6.3	$4.5_{\pm 0.3}$	1.6E13	1.1E12 6.3	$5.0_{\pm 0.3}$	2.9E13	6.4E11 11.7	$6.5_{\pm 0.5}$	4.6
	sc	1.4E12	2.4E10 11.7	$8.3_{\pm 0.4}$	6.1E12	9.3E10 9.0	$6.0_{\pm 0.4}$	1.6E13	1.1E11 11.7	$6.0_{\pm 0.4}$	2.6E13	2.0E11 1.7	$6.3_{\pm 0.3}$	6.7
C_2H_4	fc	-	-	-	-	-	-	7.8E11	- -	-	1.3E12	3.4E11 3.6	$6.4_{\pm 4.3}$	6.4
	sc	-	-	-	-	-	-	3.0E11	9.2E10 3.6	$6.5_{\pm 1.8}$	6.5E11	9.1E10 3.6	$6.0_{\pm 4.7}$	6.3
C_2H_6	fc	-	-	-	5.0E12	-	-	1.3E13	7.4E12 3.6	$13.9_{\pm 6.1}$	3.2E13	1.0E13 9.0	$14.4_{\pm 2.5}$	**14.2**

The results for four different flow rates of methane or methanol are given as the main column headings in the tables. They correspond to percentages of 0.9, 2.7, 4.5 and 7.2. For each flow rate there are three sub-columns. First, the species concentration without added oxygen ($n[O_2]= 0$), second, the minimum detected concentration on the exponential decay line for a given amount of added oxygen ($n_{min}/O_2[\%]$), and last, the amount of oxygen required to cause an order of magnitude decay of the species concentration (n decay $O_2[\%]$). Generally the decays of the C-2 hydrocarbon species were not significantly different for methane and methanol plasmas. In a similar manner to the methyl radical, exponential decays of concentration were observed for C_2H_2, C_2H_4 and C_2H_6 in methane plasmas and for CH_4, C_2H_2, C_2H_4 and C_2H_6 in methanol plasmas as the flow of oxygen was increased. The magnitude of the hydrocarbon concentration decays with $[O_2]$ can be averaged over the different percentages of methane and methanol. This average n decay $O_2[\%]$ is given in the final column of Table 6.5 and Table 6.6. It can be used to categorise the decays into two groups: (a) decays between 13 and 18 % and (b), between 4 and 7 %. Category (a) applies to CH_3 and C_2H_6 in methane under flowing conditions (Table 6.5) and CH_3, CH_4, C_2H_6 in methanol (fc Table 6.6), and also methane decays in methanol under static conditions. All other product hydrocarbon species decays fell in category (b).

These qualitative differences in decay rates can be discussed in terms of the principal production and decay processes involved. Both electron induced processes and chemical reactions influence the formation and decay of hydrocarbons. Methyl radical production is mainly due to electron impact dissociation of methane or methanol. Under flowing conditions the effective residence time of the precursor species is the time it spends in the plasma i.e. it is quite short. Hence, rather more oxygen is required to induce a measurable decay than in a static system. This leads to case (a) behaviour. Since ethane is mainly formed by methyl recombination this species is expected to behave similarly to methyl (Table 6.5 and Table 6.6, fc). In contrast under static conditions methyl decays rapidly with oxygen addition (lower average n decay $O_2[\%]$ in Table 6.5 and Table 6.6). The corresponding decays for ethane (sc) were not available because the ethane concentration was below the detection limit. The methane appearing as a product in methanol plasmas (fc and sc) also confirms the case (a) behaviour. All other species are controlled mainly by chemical reactions which can take place outside as well as inside the plasma and they fall into category (b).

6.4.2   Carbon monoxide and carbon dioxide

The concentrations of both CO and CO_2 increase as the oxygen flow increases (Figure 6.25 and Figure 6.26), although for CO formation a maximum is reached at much lower oxygen content. Both molecules are believed to represent the final products for the reaction chain when adding oxygen, just like in atmospheric chemistry. For CO, the largest concentrations appear using the methane precursor under static conditions while for CO_2 formation it is the methanol precursor that forms the most CO_2. The main chemical species leading to the formation of CO is believed to be formaldehyde which is produced by the reaction of O or OH with the methyl radical. It is interesting to note that methanol plasmas contain more formaldehyde than methane plasmas under both flowing and static conditions (Figure 6.13 to Figure 6.16).

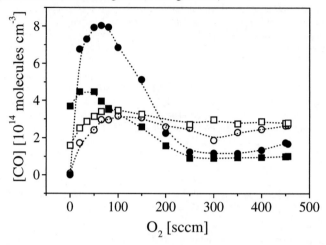

Figure 6.25   *Variation of the carbon monoxide concentration with oxygen flow rate measured by TDLAS. The flow rates of methanol or methane were 40 sccm. Legend:* ○ *- CH_4, fc;* □ *- CH_3OH, fc;* ● *- CH_4, sc;* ■ *- CH_3OH, sc. [44]*

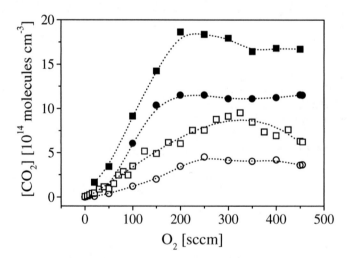

Figure 6.26   *Variation of the carbon dioxide concentration with oxygen flow rate measured by TDLAS. The flow rates of methanol or methane were 40 sccm Legend:* ○ *- CH_4, fc;* □ *- CH_3OH, fc;* ● *- CH_4, sc;* ■ *- CH_3OH, sc. [44]*

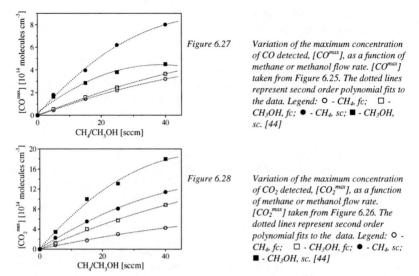

Figure 6.27 *Variation of the maximum concentration of CO detected, [CO^max], as a function of methane or methanol flow rate. [CO^max] taken from Figure 6.25. The dotted lines represent second order polynomial fits to the data. Legend: O - CH₄, fc; ☐ - CH₃OH, fc; ● - CH₄, sc; ■ - CH₃OH, sc. [44]*

Figure 6.28 *Variation of the maximum concentration of CO₂ detected, [CO₂^max], as a function of methane or methanol flow rate. [CO₂^max] taken from Figure 6.26. The dotted lines represent second order polynomial fits to the data. Legend: O - CH₄, fc; ☐ - CH₃OH, fc; ● - CH₄, sc; ■ - CH₃OH, sc. [44]*

The data in Figure 6.25 and Figure 6.26 were recorded for 40 sccm of CH_4 or CH_3OH. The variation of the maximum CO or CO_2 concentration with methane or methanol was investigated (Figure 6.27 and Figure 6.28). As expected the concentrations of both oxides increased as the hydrocarbon precursor concentration increased. Most data were found to fit a second order polynomial function as shown in the figures.

6.4.3   Formaldehyde and formic acid

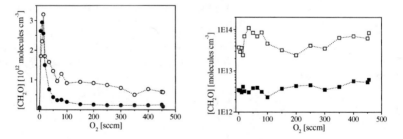

Figure 6.29 *Variation of the formaldehyde concentration with oxygen flow rate at 40 sccm hydrocarbon flow rate measured by TDLAS. Legend: O - CH₄, fc; ☐ - CH₃OH, fc; ● - CH₄, sc; ■ - CH₃OH, sc.*

Formaldehyde was detected in both types of hydrocarbon plasmas containing oxygen (for both fc and sc), see Figure 6.29, although it was always considerably higher in methanol than in methane plasmas and for flowing rather than static conditions. The left hand figure shows the results of the fc and sc methane-containing plasma and the right hand figure those of the methanol-containing plasma.

The one order of magnitude higher formaldehyde concentration in the fc methanol-containing plasma is remarkable because it differs from the conditions inside the methane-containing plasma. Higher formaldehyde concentrations when using fc support the idea of the intermediate formation of this molecule. While dependencies of molecular concentrations inside the different plasmas based on the respective hydrocarbon are mainly similar, those of formaldehyde are different. As shown in Figure 6.29 left panel a maximum of the formaldehyde concentration was found at rather low oxygen contents. With increasing amount of oxygen the value decreases and becomes nearly constant.

In oxygen containing plasmas the reaction of the oxygen atom with the methyl radical, which is very fast ($k = 1.4 \times 10^{-10}$ cm^3 molecule^{-1} s^{-1} [155]), is one of the main reactions leading to lower C-2 hydrocarbon concentrations. This reaction leads to the formation of formaldehyde although there may be other sources of formaldehyde in methanol plasmas with or without added oxygen.

Formic acid was only detected in methanol plasmas (Figure 6.14 and Figure 6.16) and was always lower in concentration than formaldehyde. The formic acid concentration was found to be independent of the different hydrocarbon flows within the error limits. Its concentration behaviour roughly followed that of formaldehyde suggesting that it was produced from CH_2O. Generally it has been found only at $O_2 > 50$ sccm or higher oxygen flows. Formic acid was not observed in methane plasmas but this may just have been a result of detection sensitivity of about 2×10^{11} molecules cm^{-3}.

6.4.4   Water

Because H_2O is the major product of both types of hydrocarbon discharge when they contain hydrogen and oxygen, its concentration is mainly governed by the ratio of hydrogen to oxygen. Figure 6.30 and Figure 6.31 show the detailed behaviour of the water concentration for fc and sc, respectively. The results of methane-containing plasmas are shown in the left hand figures and those of methanol-containing plasmas in the right hand figures. Two curves are shown, that for the highest and the lowest hydrocarbon content. The overall profile of the curves presented in the same figure was reproduced for other flows of both methane and methanol. The two curves of 25 sccm and 15 sccm are neglected because they lie between the shown curves so that the limiting cases of 40 sccm and 5 sccm encompass the complete behaviour.

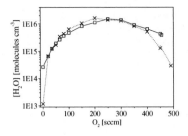 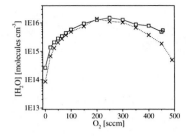

Figure 6.30   *Concentrations of H₂O measured by TDLAS in a methane plasma (left panel) and in a methanol plasma (right panel) as a function of molecular oxygen flow rate (fc). The increase in the O₂ flow rate was balanced by a reduction in the H₂ flow rate. Hydrocarbon flow rates:* □ *- 40 sccm;* ✕ *- 5 sccm.*

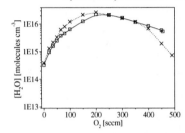 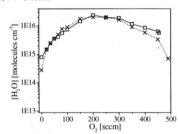

Figure 6.31   *Concentrations of H₂O measured by TDLAS in a methane plasma (left panel) [44] and in a methanol plasma (right panel) as a function of molecular oxygen flow rate (sc). The increase in the O₂ flow rate was balanced by a reduction in the H₂ flow rate. Hydrocarbon flow rates:* □ *- 40 sccm;* ✕ *- 5 sccm.*

The basic experimental result for the water concentration is that it rises steeply to a maximum at $H_2:O_2$ ratios of about 1:1. Furthermore, the water concentration increases strongly already with small amounts of oxygen. However, as can be seen in Figure 6.30 and Figure 6.31 the curves for different flow rates of hydrocarbon are not superimposable. The water concentration for the larger flow of methane (40 sccm) is lower when $O_2 < 250$ sccm and higher when $O_2 > 250$ sccm than observed for the lesser flow rate of methane (5 sccm). This observation implies that more of the added O_2 is consumed by the methane at low oxygen contents thus producing less water. The same tendency was found in methane-containing plasmas under fc and sc, but more pronounced under sc. This effect could not be observed for the methanol case. Generally, when using high oxygen contents there is always a deficit of hydrogen, which is compensated for by reactions of the hydrocarbon precursor. The larger hydrocarbon concentration must be able to create more hydrogen.

Water has been found even in plasmas without additional oxygen. In case of using methanol the water molecules can be produced by the OH radical, what is expected as dissociation product of methanol. Especially under sc, see Figure 6.31, a slight dependence on the amount of methanol has been found for the first data point. In contrast to these results the water concentration using methane (sc) was found to be independent on the amount of methane. These water concentrations show reactor system leakages. On the other hand the used reactor is constructed for investigations comparable to technological purposes thus it is no high vacuum system and the reactor walls are not heated externally. A certain latent amount of water is stored inside aluminium walls and can leave them by outgassing depend on the discharge history.

6.4.5   Production of methane

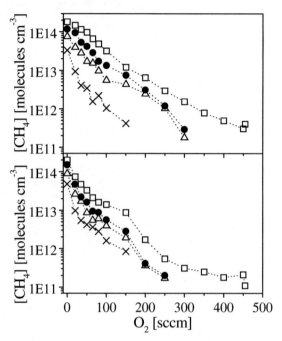

Figure 6.32   Variation of methane concentration with oxygen contents using methanol-containing source gas. Upper panel - fc, lower panel - sc. Flow rates: □ - 40 sccm; ● - 25 sccm; △ - 15 sccm; ✕ - 5 sccm.

For methanol plasmas methane is a product molecule and it appears at approximately the same concentration as carbon monoxide in the absence of oxygen, and decreases exponentially as the oxygen flow is increased (Figure 6.14 and Figure 6.16). Figure 6.32 shows the methane concentration of the four different methanol flow rates in the fc plasma (upper panel) and the sc plasma (lower panel). The highest methane concentration was found with the highest methanol content and without added oxygen. Further there is no significant difference between fc and sc.

6.5 Mass balances

Furthermore, the mass balances with respect to carbon, hydrogen and oxygen are presented for sc. The carbon mass balance for H_2-O_2-Ar plasmas containing 7.2 % of methane or methanol can be seen in Figure 6.33.

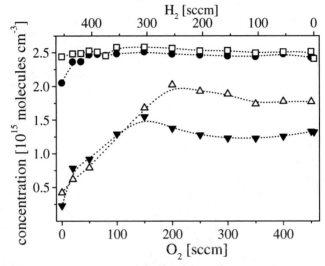

Figure 6.33   *Carbon mass balance in methane and methanol plasmas (hydrocarbon content 7.2 %, sc) as a function of the oxygen and hydrogen concentration. The concentrations of methane (●) and methanol (□) are the amounts consumed in the plasmas. The corresponding total amount of carbon-containing molecules measured by TDLAS is represented by (▼) for methane and (△) for methanol.*

It has been found by comparing the concentrations of CO and CO_2 formed with the concentration of hydrocarbons removed that about 50 % of the available methane and up to 80 % of the available methanol appear as CO and CO_2. The increasing degree of dissociation with oxygen amount results mainly in the production of CO and CO_2.

At low oxygen content the measured proportion of carbon is much smaller than that using medium or higher oxygen content. The sum of the concentrations of the produced hydrocarbons inside the methane-containing plasma is less than 3×10^{14} cm^{-3}. In the methanol-containing plasma this value is even lower, about 7×10^{13} cm^{-3}, thus both values can be neglected compared with the concentrations of CO and CO_2 except the discharge without additional oxygen. In the H_2-Ar-CH_4 plasma only the hydrocarbons contribute to the mass balance. But hydrocarbons, formaldehyde and carbon monoxide have been measured in the H_2-Ar-CH_3OH plasma, whereas the carbon monoxide concentration already exceeded that of the other species.

The most free carbon for the production of molecules, complex dust particles or deposition effects was found at no or low oxygen amounts. It corresponds to studies of the carbon thin film deposition inside plasmas having H_2-O_2-CH_4 gas mixtures [e.g. [156]]. Similar behaviour has also been found in H_2-O_2-Ar-CH_3OH gas mixtures. In fact, the difference between the consumed and the measured carbon has the maximum in the H_2-Ar-CH_3OH mixture, but already small amounts of oxygen lead to the conversion of carbon into formaldehyde and carbon monoxide. When further using higher amounts of oxygen the production of carbon dioxide is very effective.

In addition to the carbon mass balance also the hydrogen and oxygen balances have been of interest. In case of hydrogen and oxygen the degree of dissociation of the source gas molecules is unknown thus the amount of introduced source gas is compared with the measured molecule amounts. The results of the hydrogen and oxygen mass balances can be seen in Figure 6.34 and Figure 6.35, respectively. Left panels show the conditions in H_2-O_2-Ar-CH_4 plasmas and right panels show those in H_2-Ar-O_2-CH_3OH plasmas.

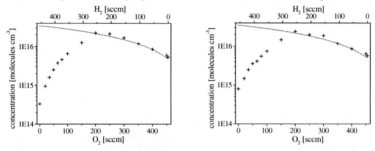

Figure 6.34   *Hydrogen mass balance of the sc methane-containing discharge (left panel) and the methanol-containing discharge (right panel). The dotted line indicates the amount of added H_2 represented by the precursor gas mixture (hydrocarbon content 7.2 %). The symbols + show the H_2 concentration found in water.*

Inside both types of plasmas the measured hydrogen concentration is dominated by the amount of water, see Figure 6.34. The further measured molecules can be neglected because of their very low concentrations compared to that of water. At oxygen flows of 200 sccm and higher the respective low hydrogen amount is converted to water. Using very high amounts of hydrogen seems to have a high potential for further free or bound hydrogen.

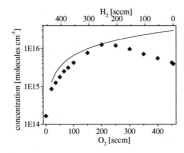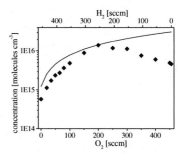

Figure 6.35   Oxygen mass balance of the sc methane-containing discharge (left panel) and the methanol-containing discharge (right panel). The solid line indicates the amount of added O_2 represented by precursor gas mixture (hydrocarbon content 7.2 %). The symbols ◆ show the O_2 concentration found in water, carbon monoxide and carbon dioxide.

In fact, the measured oxygen amount, see Figure 6.35, is also strongly influenced by the behaviour of the water concentration, but the concentrations of carbon monoxide and carbon dioxide take part, as well. Therefore the sum of the mentioned species is shown and further oxygen containing molecule concentrations were neglected. Only at 200 sccm of oxygen the introduced oxygen amount is completely converted to water, carbon monoxide and carbon dioxide. Using higher or lower oxygen amounts always a difference between measured and introduced values appears.

Only at the oxygen to hydrogen ratio of 200 to 255 the mass balance of hydrogen and oxygen is completed, mainly by water but also by carbon monoxide and carbon dioxide. Other molecules containing carbon, hydrogen and oxygen can be generally expected. It is a future challenge to measure the higher-chain species and dust particles.

7   Model Calculations

Extensive experimental investigations were accompanied by model calculations of the kinetic processes in the plasma.

High quality models include a plasma model, a plasma chemistry model and the surface model. Each of these modules can be described by equations that illustrate the temporal and spatial variation of characteristics. Plasma models mainly describe the behaviour of the charged particles using transport and kinetic equations as well as Maxwell's equations. Plasma chemistry models specify the chemical reactions of neutrals using transport and kinetic equations, and surface models shows the deposition or etching processes at walls. In general all components are coupled with each other. Often main restriction is the lack of the respective, e.g., cross sections or rate coefficients, so that reasonable simplifications are necessary. Often only two modules are treated or one module is used with some general assumptions adapted to the plasma used.

Current plasma chemical models mainly include volume reactions between neutral species and neutral-ion processes and their surface interactions leading to the layer formation. The number of known rate coefficients is rather limited so that reactions are either partly neglected or approximately estimated. Also some approaches involving the Boltzmann equation are used. All these methods a high effort in developing time and computational equipment is common.

There have been several reports in the literature of modelling the chemistry of hydrocarbon plasmas under different conditions, but mainly focused on rf plasmas [1,2,61,157,158,159,160], hot filament reactors [13,161] or the plasma jet [162]. The need of understanding the carbon layer depositions has led to several calculations. Some of the authors were able to develop a simplified reaction scheme for the volume processes, also in combination with the growth mechanism at surfaces. The calculated molecular concentrations are only partly compared with measurements done by e.g. mass spectrometry [2]. In the majority of examples methane is the precursor. Cases where other hydrocarbons, like e.g. C_2H_4 or C_2H_6, are used or where reactions with added oxygen are a possibility are few and far between.

On the other hand, for hydrocarbon-containing microwave plasmas no comparable theoretical studies have been performed so far. To overcome this situation the set of kinetic processes acquired for the other plasma types can be partly used, but should be new validated for microwave plasmas. Especially the rate coefficients of electron impact processes can vary because of different electric field strengths, pressures, temperatures and mixture compositions and thus of a different energy distribution function of the electrons.

The aim of the present model calculations is to improve the physical understanding of the trends of methyl and associated species concentrations. The calculated species concentrations of the several molecules are compared with corresponding experimental results in order to obtain better information about the principal reaction paths in these complex chemically reactive plasmas.

In preparation of the present task it is advantageous to have assembled a flexible computational toolbox. Therefore, electron impact dissociation rate coefficients and the rate coefficients of chemical reactions of neutral molecules were taken from literature. Collections of previously used chemical reactions were used. With respect to the present plasma conditions this reaction mechanism is used to generate a set of rate equations. This set of usual differential equations is solved by an integration technique. Finally the results are used to compare the experimental and theoretical molecular concentrations and to analyse the main reaction paths.

7.1   Main aspects of the model

In the present work model computations are presented for H_2-O_2-Ar microwave plasmas containing methane. Therefore a simple model includes dissociative electron-neutral collisions and neutral-neutral collisions as chemical reactions. The plasma zone with a size of about 170 cm^3 is assumed to be spatially uniform. The power density is about 17 W cm^{-2}. The distances to metallic walls were small compared with the mean free paths of the molecules. Mean free paths of about 100 μm were estimated for molecule-molecule collisions and about 400 μm for electron-molecule collisions. This estimation was done using the hard sphere assumption of the molecules for calculating the cross sections. The distance to the microwave window which consists of quartz glass is much shorter, but all neutral species are assumed to be reflected there without chemical or energetic changes. In order to reduce the number of unknown or poorly known parameters a static cell model was assumed and the results are compared with the experimental static conditions (sc). This means in the plasma is no gas flow, and temperature and volume are set constant. Additional steps required for mass transport and diffusion could therefore be excluded. In these model calculations ionic processes are not treated and the model does not include an electron rate equation. In accordance with experiments in the afterglow the electron density of 1×10^{12} cm^{-3} has been assumed.

The rate coefficients of the chemical reactions were taken from the combustion literature [155] and from previous investigations using different plasma types [161,163,164]. The data from literature are often expressed in terms of the Arrhenius formalism. The reaction rate coefficient K is dependent on the temperature T and the activation energy E_A and is calculated as

$$K = K_0 T^n \exp\left(\frac{E_A}{kT}\right) \qquad (7.25),$$

K_0 is the so-called preexponential factor.

The electron impact dissociation rate coefficients for H_2 and CH_4 dependent on the reduced field strength are shown in Figure 7.36.

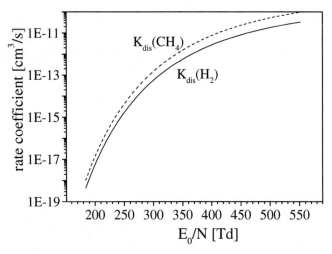

Figure 7.36   Calculated electron impact dissociation rate coefficients for methane and hydrogen dependent on the reduced field strength. [165]

The electron impact dissociation rate coefficients of H_2 and CH_4 have been determined using the time-dependent Boltzmann equation with respect to the microwave frequency of 2.45 GHz, the discharge pressure of 1.5 mbar the $H_2/Ar/CH_4$ mixture of 82/10.8/7.2 and the gas temperature of 1000 K [165]. Beside the selected field parameters and the gas composition also atomic data, as cross sections, masses of involved species and energy losses due to inelastic collisions were used as input for this equation. Therefore, cross sections for electron-molecule (H_2, CH_4) and electron-argon collisions were taken from the established literature [166,167,168]. The field parameter of the ac field is

$$E_{ac} = E_0 \cos \omega t \qquad (7.26)$$

where E_0 is the amplitude of the actual ac field. Integration of the field strength over the period results in an effective electric field strength. In case of ac discharges it should be used instead of the dc field strength. The Boltzmann equation has been solved up to the establishing of the periodic state. The rate coefficients are averaged over the period and given as a function of E_0/N, as shown in Figure 7.36.

Basic values like the cross section, temperature coefficient and activation energy are converted into useful coefficients.

Calculations of the reaction mechanism were carried out for the planar reactor configuration, described in chapter 3.2. The reaction kinetic model is based on a set of rate equations for 24 components and about 125 of their reactions have been taken into account. The rate coefficients for the various chemical reactions and the electron impact dissociation processes have mainly been taken from literature [1,61,99,161,155,163,164]. Reactions were included consisting of electron impact steps and chemical reactions. The used electron impact dissociation reactions are presented in Table 7.7.

no.	reaction		rate coefficient [cm^{3}s^{-1}]	reference
e 1	$e^- + H_2$	$\rightarrow H + H + e^-$	4×10^{-14}	$K(E_0/N)$ []
e 2	$e^- + CH_4$	$\rightarrow CH_3 + H + e^-$	9×10^{-14}	$K(E_0/N)$ [165], [1]
e 3	$e^- + CH_4$	$\rightarrow CH_2 + H_2 + e^-$	1×10^{-14}	$K(E_0/N)$ [165], [1]
e 4	$e^- + CH_4$	$\rightarrow CH + H_2 + H + e^-$	7×10^{-15}	$K(E_0/N)$ [165], [1]
e 5	$e^- + C_2H_6$	$\rightarrow C_2H_5 + H + e^-$	1×10^{-13}	estimated
e 6	$e^- + C_2H_6$	$\rightarrow 2\,CH_3 + e^-$	5×10^{-14}	estimated
e 7	$e^- + C_2H_5$	$\rightarrow C_2H_4 + H + e^-$	1×10^{-13}	estimated
e 8	$e^- + C_2H_4$	$\rightarrow C_2H_2 + H_2 + e^-$	1×10^{-13}	estimated
e 9	$e^- + C_2H_4$	$\rightarrow C_2H_3 + H + e^-$	1×10^{-13}	estimated
e 10	$e^- + C_2H_2$	$\rightarrow C_2H + H + e^-$	1×10^{-13}	estimated
e 11	$e^- + O_2$	$\rightarrow O + O + e^-$	2×10^{-14}	estimated
e 12	$e^- + HO_2$	$\rightarrow OH + O + e^-$	1×10^{-12}	estimated
e 13	$e^- + H_2O$	$\rightarrow OH + H + e^-$	1×10^{-14}	estimated
e 14	$e^- + CH_2O$	$\rightarrow HCO + H + e^-$	1×10^{-11}	estimated
e 15	$e^- + CH_2O$	$\rightarrow CO + H_2 + e^-$	6×10^{-10}	estimated
e 16	$e^- + CO_2$	$\rightarrow CO + O + e^-$	9×10^{-11}	estimated

Table 7.7   Electron - molecule collisions and their dissociation rate coefficients

The dissociation rate coefficients of H_2 and CH_4 have been obtained using the rate coefficients for an E_0/N of about 290 Td. The ratio between the different channels of the methane dissociation was taken from Kline et al. [1]. The other rate coefficients of the electron-molecule collisions were chosen with respect to the value of the methane dissociation assuming a related order of magnitude.

The neutral chemical reactions relevant for H_2-O_2-Ar-CH_4 plasmas and their rate coefficients given for 1000 K and 1.5 mbar are reported in Table 7.8.

no.	reaction		rate coefficient [cm^{3}s^{-1}; cm^{6}s^{-1}]	reference
c 1	$H_2 + H_2$	$\rightarrow 2\,H + H_2$	1.19×10^{-11}	[155]
c 2	$H_2 + Ar$	$\rightarrow 2\,H + Ar$	2.94×10^{-14}	[155]
c 3	$H + H + M$	$\rightarrow H_2 + M$	2.35×10^{-25}	[161]
c 4	$CH_4 + H$	$\rightarrow CH_3 + H_2$	1.47×10^{-11}	[155]
c 5	$CH_4 + Ar$	$\rightarrow CH_3 + H_2$	1.09×10^{-8}	[155]
c 6	$CH_4 + CH_4$	$\rightarrow CH_3 + H + CH_4$	1.14×10^{-7}	[155]
c 7	$CH + H$	$\rightarrow C + H_2$	2.49×10^{-10}	[161]

to be continued

no.	reaction	rate coefficient [cm^3s^{-1}; cm^6s^{-1}]	reference
c 8	$CH + CH_4 \rightarrow C_2H_4 + H$	9.3×10^{-11}	[99][a]
c 9	$CH_2 + H \rightarrow CH + H_2$	6.64×10^{-11}	[161]
c 10	$CH_2 + CH_4 \rightarrow 2\ CH_3$	1.34×10^{-11}	[163]
c 11	$CH_3 + H + M \rightarrow CH_4 + M$	1.95×10^{-28}	[99][a]
c 12	$CH_3 + H \rightarrow CH_2 + H_2$	1.40×10^{-10}	[161]
c 13	$CH_3 + H_2 \rightarrow CH_4 + H$	1.35×10^{-12}	[161]
c 14	$CH_3 + H_2 \rightarrow CH_2 + H_2 + H$	5.05×10^{-10}	[161]
c 15	$CH_3 + CH_3 + M \rightarrow C_2H_6 + M$	4.6×10^{-27}	[164]
c 16	$CH_3 + CH_3 \rightarrow C_2H_5 + H$	9.12×10^{-12}	[161]
c 17	$C_2H_6 + H \rightarrow C_2H_5 + H_2$	2.18×10^{-11}	[161]
c 18	$C_2H_6 + CH_3 \rightarrow C_2H_5 + CH_4$	6.02×10^{-13}	[161]
c 19	$C_2H_5 + H_2 \rightarrow C_2H_6 + H$	3×10^{-15}	estimated
c 20	$C_2H_5 + H \rightarrow C_2H_4 + H_2$	4.98×10^{-11}	[161]
c 21	$C_2H_5 + H \rightarrow 2\ CH_3$	6×10^{-11}	[155]
c 22	$C_2H_4 + H \rightarrow C_2H_3 + H_2$	4.42×10^{-11}	[161]
c 23	$C_2H_4 + H_2 \rightarrow C_2H_2 + H_2 + H_2$	2.32×10^{-8}	[161]
c 24	$C_2H_4 + H_2 \rightarrow C_2H_3 + H_2 + H$	9.72×10^{-9}	[161]
c 25	$C_2H_4 + H_2 \rightarrow C_2H_5 + H$	5×10^{-15}	estimated
c 26	$C_2H_3 + H \rightarrow C_2H_2 + H_2$	4.98×10^{-11}	[161]
c 27	$C_2H_3 + H_2 \rightarrow C_2H_4 + H$	9.6×10^{-17}	[99][a]
c 28	$C_2H_2 + H + M \rightarrow C_2H_3 + M$	2.08×10^{-28}	[161]
c 29	$C_2H_2 + H \rightarrow C_2H + H_2$	2.47×10^{-11}	[155]
c 30	$C_2H_2 + H_2 \rightarrow C_2H_3 + H$	1.5×10^{-17}	estimated
c 31	$C_2H + H_2 \rightarrow C_2H_2 + H$	1.16×10^{-11}	[161]
c 32	$CH_2 + CH_3 \rightarrow C_2H_4 + H$	8.30×10^{-11}	[161]
c 33	$CH_2 + CH_2 \rightarrow C_2H_2 + H_2$	1.92×10^{-11}	[163]
c 34	$CH_2 + CH_2 \rightarrow C_2H_2 + H + H$	1.73×10^{-10}	[163]
c 35	$C_2H_3 + C_2H_5 \rightarrow C_2H_2 + C_2H_6$	7.79×10^{-13}	[163]
c 36	$C + CH_3 \rightarrow C_2H_2 + H$	8.30×10^{-11}	[161]
c 37	$C_2H_4 + CH_3 \rightarrow C_2H_3 + CH_4$	3.98×10^{-13}	[161,163]
c 38	$C_2H_3 + CH_2 \rightarrow C_2H_2 + CH_3$	2.99×10^{-11}	[163]

to be continued

no.	reaction	rate coefficient [cm^{3}s^{-1}; cm^{6}s^{-1}]	reference
c 39	$C_2H_3 + CH_3 \rightarrow C_2H_2 + CH_4$	6.48×10^{-13}	[163]
c 40	$C_2H_3 + C_2H_3 \rightarrow C_2H_2 + C_2H_4$	1.59×10^{-12}	[163]
c 41	$C_2H_5 + CH_2 \rightarrow C_2H_4 + CH_3$	2.99×10^{-11}	[163]
c 42	$C_2H_5 + CH_3 \rightarrow C_2H_4 + CH_4$	7.9×10^{-11}	[164]
c 43	$C_2H_3 + C_2H_5 \rightarrow 2\ C_2H_4$	7.97×10^{-13}	[163]
c 44	$C_2H_5 + C_2H_4 \rightarrow C_2H_6 + C_2H_3$	8.46×10^{-13}	[163]
c 45	$C_2H_5 + C_2H_5 \rightarrow C_2H_4 + C_2H_6$	2.4×10^{-12}	[155]
c 46	$O_2 + H \rightarrow OH + O$	1.87×10^{-10}	[161]
c 47	$O_2 + H + M \rightarrow HO_2 + M$	1.7×10^{-24}	[161]
c 48	$O + H_2 \rightarrow OH + H$	5.91×10^{-12}	[161]
c 49	$OH + H_2 \rightarrow H_2O + H$	6.54×10^{-12}	[155]
c 50	$HO_2 + H \rightarrow 2\ OH$	2.37×10^{-10}	[161]
c 51	$HO_2 + H \rightarrow H_2 + O_2$	4.01×10^{-11}	[161]
c 52	$HO_2 + O \rightarrow OH + O_2$	5.3×10^{-11}	[155]
c 53	$HO_2 + H \rightarrow H_2O + O$	4.58×10^{-11}	[155]
c 54	$HO_2 + OH \rightarrow H_2O + O_2$	4.68×10^{-11}	[155]
c 55	$H_2O + H \rightarrow OH + H_2$	1.87×10^{-11}	[155]
c 56	$CH_4 + O \rightarrow CH_3 + OH$	4.26×10^{-11}	[161]
c 57	$CH_4 + OH \rightarrow CH_3 + H_2O$	5.43×10^{-12}	[161]
c 58	$CH_3 + OH \rightarrow CH_2O + H_2$	1.33×10^{-11}	[161]
c 59	$CH_3 + O \rightarrow CH_2O + H$	1.4×10^{-10}	[161]
c 60	$CH_3 + O_2 \rightarrow CH_2O + OH$	8.04×10^{-13}	[161]
c 61	$CH_3 + OH \rightarrow CH_2 + H_2O$	4.06×10^{-12}	[161]
c 62	$CH_3 + OH \rightarrow CH_4 + O$	1×10^{-10}	estimated
c 63	$CH_3 + HO_2 \rightarrow CH_4 + O_2$	1×10^{-10}	estimated
c 64	$C_2H_6 + OH \rightarrow C_2H_5 + H_2O$	1.77×10^{-11}	[161]
c 65	$C_2H_6 + O \rightarrow C_2H_5 + OH$	3.92×10^{-11}	[155]
c 66	$C_2H_6 + O_2 \rightarrow C_2H_5 + HO_2$	7.35×10^{-12}	[155]
c 67	$C_2H_5 + O \rightarrow CH_2O + CH_3$	1.87×10^{-11}	[155]
c 68	$C_2H_5 + O_2 \rightarrow C_2H_4 + HO_2$	1.7×10^{-14}	[155]
c 69	$C_2H_4 + OH \rightarrow C_2H_3 + H_2O$	2.5×10^{-11}	[155]

to be continued

no.	reaction		rate coefficient [cm^3s^{-1}; cm^6s^{-1}]	reference
c 70	$C_2H_4 + O$	$\rightarrow CH_2O + CH_2$	6×10^{-12}	[155]
c 71	$C_2H_4 + O$	$\rightarrow HCO + CH_3$	3.5×10^{-12}	[155]
c 72	$C_2H_3 + O_2$	$\rightarrow CH_2O + HCO$	6.75×10^{-12}	[161]
c 73	$C_2H_3 + O$	$\rightarrow C_2H_2 + OH$	1.67×10^{-11}	[155]
c 74	$C_2H_3 + O$	$\rightarrow CO + CH_3$	1.67×10^{-11}	[155]
c 75	$C_2H_3 + O$	$\rightarrow HCO + CH_2$	1.67×10^{-11}	[155]
c 76	$C_2H_3 + OH$	$\rightarrow C_2H_2 + H_2O$	6.64×10^{-12}	[161]
c 77	$C_2H_2 + O$	$\rightarrow CO + CH_2$	4.41×10^{-12}	[155]
c 78	$C_2H_2 + OH$	$\rightarrow C_2H + H_2O$	2.65×10^{-11}	[161]
c 79	$C_2H + O$	$\rightarrow CO + CH$	1.7×10^{-11}	[155]
c 80	$C_2H + O_2$	$\rightarrow CO_2 + CH$	7.5×10^{-12}	[155]
c 81	$C_2H + O_2$	$\rightarrow 2\ CO + H$	7.5×10^{-12}	[155]
c 82	$C_2H + O_2$	$\rightarrow CO + HCO$	7.5×10^{-12}	[155]
c 83	$CH + O$	$\rightarrow CO + H$	6.6×10^{-11}	[155]
c 84	$C + O_2$	$\rightarrow CO + O$	3.32×10^{-11}	[161]
c 85	$C + OH$	$\rightarrow CO + H$	8.3×10^{-11}	[161]
c 86	$CH_2 + O$	$\rightarrow CO + H_2$	8×10^{-11}	[155]
c 87	$CH_2 + O$	$\rightarrow CO + 2\ H$	1.2×10^{-10}	[155]
c 89	$CH_2 + OH$	$\rightarrow CH_2O + H$	4.15×10^{-11}	[161]
c 90	$CH_2 + OH$	$\rightarrow CH + H_2O$	1.62×10^{-11}	[161]
c 91	$CH_2 + O_2$	$\rightarrow CO + OH + H$	2.66×10^{-12}	[161]
c 92	$CH_2O + OH$	$\rightarrow HCO + H_2O$	3.02×10^{-11}	[161]
c 93	$CH_2O + O$	$\rightarrow HCO + OH$	3.08×10^{-11}	[155]
c 94	$CH_2O + H$	$\rightarrow HCO + H_2$	1.36×10^{-11}	[161]
c 95	$CH_2O + CH_3$	$\rightarrow CH_4 + HCO$	2.46×10^{-14}	[161]
c 96	$CH_2O + H_2$	$\rightarrow HCO + H + H_2$	5.13×10^{-9}	[161]
c 97	$CH_2O + H_2$	$\rightarrow CO + 2\ H_2$	1.20×10^{-9}	[161]
c 98	$CH_2O + H_2O$	$\rightarrow CH_3 + HO_2$	8×10^{-15}	estimated
c 99	$HCO + H$	$\rightarrow CO + H_2$	3.32×10^{-10}	[161]
c 100	$HCO + OH$	$\rightarrow CO + H_2O$	1.66×10^{-10}	[161]
c 101	$HCO + O$	$\rightarrow CO + OH$	5×10^{-11}	[155]

to be continued

no.	reaction		rate coefficient [cm^{3}s^{-1}; cm^{6}s^{-1}]	reference
c 102	HCO + O	$\rightarrow$ CO$_2$ + H	5 x 10^{-11}	[155]
c 103	HCO + H$_2$	$\rightarrow$ CH + O + H$_2$	2.49 x 10^{-13}	[161]
c 104	HCO + HCO	$\rightarrow$ CH$_2$O + CO	5 x 10^{-11}	[155]
c 105	HCO + H$_2$O	$\rightarrow$ CH$_3$ + O$_2$	7 x 10^{-14}	estimated
c 106	CO + O + M	$\rightarrow$ CO$_2$ + M	1.36 x 10^{-26}	[161]
c 107	CO + OH	$\rightarrow$ CO$_2$ + H	2.32 x 10^{-13}	[161]
c 108	CO + H$_2$O	$\rightarrow$ HCO + OH	3 x 10^{-14}	estimated

Table 7.8   Chemical reactions and their rate coefficients at 1000 K (a - assuming $K_{325K} = K_{1000K}$)

Generally every chemical equilibrium reaction can be written like the symbolic equation

$$\gamma_1 A_1 + \gamma_2 A_2 + ... + \gamma_n A_n = \gamma_1' A_1' + \gamma_2' A_2' + ... + \gamma_n' A_n' \qquad (7.27)$$

whereas A_i and A_i' with $i = 1,...,n$ are the reagents and products, respectively. The values of γ_i and γ_i' denote their stoichiometric coefficients.

The time behaviour of the concentration [C_k] of the k-th species can be expressed by the rate equation

$$\frac{d[C_k]}{dt} = \sum_{j=1}^{m} \pm \gamma_{C_k,j} \; K_j \left[A_{j,1}(t) \right]^{\gamma_{j,1}} \left[A_{j,2}(t) \right]^{\gamma_{j,2}} \cdots \left[A_{j,n}(t) \right]^{\gamma_{j,n}} \qquad (7.28)$$

where K_j is the respective reaction rate coefficient of the j-th reaction with $j = 1,...,m$ involved reactions. The positive or negative sign indicates whether the species C_k is produced or consumed.

In the plasma under the low pressure conditions mainly binary collision occur. In this special case of only reaction (c4) in Table 4.1 equation (7.28) simplifies for CH$_4$ to

$$\frac{d[CH_4]}{dt} = -K[CH_4](t)[H](t) \qquad (7.29)$$

and for CH$_3$ to

$$\frac{d[CH_3]}{dt} = K[CH_4](t)[H](t) \qquad (7.30)$$

The rate equations for all species form a system of ordinary differential equations. In order to take the full advantage of external resources the programming package KINEL [169,170] was used to generate the rate equations and to solve the system and to analyse the importance of single reactions. Starting from different mixture compositions the concentrations of the species under static conditions have been determined by means of a time-dependent relaxation procedure. Therefore, the system of equation has been solved numerically for constant temperature and volume by Gear's method [171]. Static condition is reached if all species concentrations become time-independent. It has been achieved after typically 300 s of the relaxation process.

KINEL is the module for reaction kinetics integrated in a package for computational modelling of two dimensional reacting flows including chemical kinetics. Because for the first approach no transport phenomena are considered only the module for the reaction kinetics was used. When the modelling was started KINEL provided a better handling of many species and their reactions than others. Also other, sometimes much more sophisticated, programming packages are available today, e.g., FACSIMILE [100,172].

But, in the frame of this mainly experimentally orientated work satisfactory results of the model calculation were achieved. They are presented in the following chapter.

7.2 Results

Model calculations have been performed for the H_2-Ar-CH_4 and O_2-Ar-CH_4 microwave plasmas with a composition of 82/10.8/7.2 %, respectively. A microwave frequency of 2.45 GHz, a total pressure of 1.5 mbar and a gas temperature of 1000 K were used.

7.2.1 H_2-Ar-CH_4 system

The comparison between the calculated and measured species concentrations of the methyl radical (CH_3), ethane (C_2H_6), ethylene (C_2H_4), acetylene (C_2H_2) and methane (CH_4) is shown in Figure 7.37.

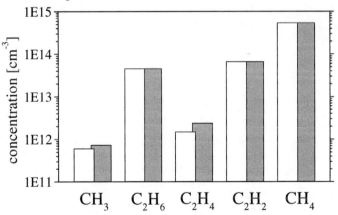

Figure 7.37   Species concentrations in a H_2-Ar-CH_4 plasma. Grey - measured by TDLAS, white out - model calculations.

The theoretical and experimental results for the species produced in the plasma agree very good within the error bars of the experiments. This agreement between the calculated and measured methyl radical concentration in accordance with further molecules is a new quality of measurements and evaluation of the modelling of kinetic processes. But also the concentrations of other transient species can be predicted from the modelling. Thus the concentration of the methylene (CH_2) radicals was calculated of the order of 3×10^{10} cm^{-3}. The further important radicals C_2H_3, C_2H_5 and C_2H should have concentrations of the order of 5×10^{11}, 9×10^{11} and 2×10^{8} cm^{-3}, respectively. Using an absorption experiment having an interaction length of the order of 15 to 20 m or even longer these predicted radicals are possible to measure by TDLAS, compare also Table 4.1 in chapter 4.1.

The main reaction paths were found by analysing the production and consumption reactions for every single species. The result is shown in Figure 7.38.

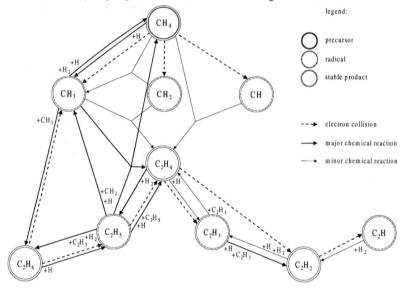

Figure 7.38   Main reaction paths of the H_2-Ar-CH_4 discharge

No comparable scheme for the hydrocarbon-containing chemistry in microwave discharges was found in the literature. The reactions are initiated by electron impact dissociation to produce the primary radicals CH_3, CH_2 and CH. A similar dissociation behaviour of methane can be found in rf discharges at very low pressures as published by Dagel et al. [2]. Not only electron induced processes but also the reactions of methane with atomic hydrogen contribute significantly to the production of methyl radicals. Ethane, ethylene and acetylene are the stable hydrocarbon products found. In investigations of rf plasmas using mass spectrometry higher hydrocarbons produced from stable and transient C-2 hydrocarbons were found, too [1,2]. A related tendency with small concentrations of higher hydrocarbons can also be assumed for microwave plasmas. Ethylene is mainly produced by chemical reactions of methane with the primary radicals CH_3, CH_2 and CH. In contrast to it the concentration of C_2H_6 is mainly balanced by the recombination of the methyl radical and reactions of C_2H_5 with hydrogen only. The electron collisions of C_2H_4 contribute to the formation of C_2H_2 but its concentration is dominated by reactions of the radical counterparts C_2H_3 and C_2H.

It is important to note the role of hydrogen in the plasma chemistry. Hydrogen atom addition or abstraction reactions with hydrocarbons including CH_3 and CH_2 contribute to the formation of ethylene and acetylene. Generally, reactions with atomic hydrogen lead to sequential reductive formation of C_2H_2 starting from C_2H_6. Reactions with molecular hydrogen also produce C_2H_6 sequentially starting from C_2H_2 although having fairly low rate coefficients. These opposite processes stabilise the whole chemistry in the model.

7.2.2 O_2-H_2-Ar-CH_4 system

The O_2-Ar-CH_4 plasma was investigated. But as soon as the methane is dissociated hydrogen appears with remarkable amounts leading to an actual O_2-H_2-Ar-CH_4 mixture. The comparison between the experimental and calculated results is shown in Figure 7.39.

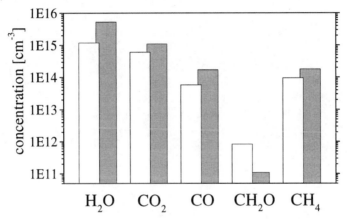

Figure 7.39   *Species concentrations in a O_2-Ar-CH_4 plasma. Grey - measured by TDLAS, white out - model calculations.*

The results show satisfactory agreement between the calculated and measured species concentrations. The largest discrepancy was found for formaldehyde about one order of magnitude. The deviations for the other molecules were found to be within half an order of magnitude.

The predicted concentrations of OH, HO_2 and HCO are 2×10^{12}, 5×10^{12} and 3×10^{12} cm^{-3}, respectively. These radicals can also be measured by TDLAS using an large interaction length. Promising spectral positions are shown in Table 4.1 in chapter 4.1.

The model calculations for the O_2-H_2-Ar-CH_4 mixtures are just at the beginning. Better agreement can be expected if the model is improved with respect to the neutral gas temperature, the dissociation rate coefficients and corresponding cross sections in oxygen-containing mixtures. This will be the subject of future studies.

Basically, the variety of molecules is extended to the stable products H_2O, CH_2O, CO and CO_2 and the radicals OH, HO_2 and HCO. By means of model calculations it was possible to predict the trends (increase or decrease) for the species CH_4, C_2H_2, C_2H_6, CH_2O and CO when oxygen was present. It became possible to predict the increasing degree of dissociation of methane and the formation of H_2O and CO_2 as more O_2 was added.

Based upon the present results by balancing the consumption and production processes a reaction scheme for O_2-H_2-Ar-CH_4 mixtures was derived.

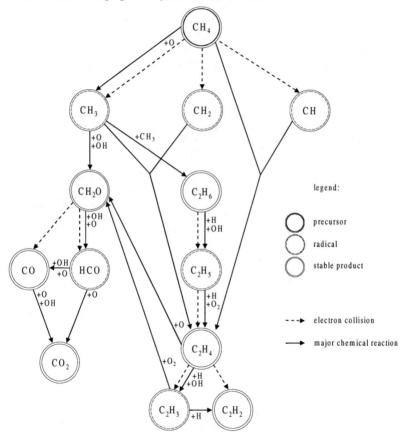

Figure 7.40   Reaction scheme showing the main contributing steps in O_2-H_2-Ar-CH_4 plasmas. Several minor steps involving oxidation of C-2 hydrocarbons have been omitted for clarity.

Figure 7.40 shows the 31 major steps linking the plasma chemistry. Additionally to the C-H reactions also reactions with oxygen become the more important the more oxygen is used in the plasma. The amount of hydrocarbons is clearly decreased in favour of the oxygen-containing molecules. Under conditions of increasing oxygen content formaldehyde is formed first and eventually CO. The final product of oxidation is CO_2. The main loss channels of C-2 hydrocarbons to formaldehyde or carbon monoxide appear from C_2H_4 or C_2H_3. Although it is stable C_2H_4 becomes also a very important intermediate molecule. Reactions of the hydrocarbons support the channel of the reduction from C_2H_6 to C_2H_2 comparable to reactions with atomic hydrogen.

Water is produced mainly by reactions of OH and HO_2 and the main consumption process is the electron impact dissociation. The experimental fact of the maximum water concentration at about 1:1 ratio of hydrogen to oxygen can also due to water formation reactions with the OH radical.

8 Methanol-Containing Plasmas

The detailed modelling of the chemical kinetics of methanol-containing plasmas was not done during this thesis. Essential data like electron collision cross sections and reaction rates are not or only poorly available. Nevertheless based on the experimental results in combination with some facts from the literature a reasonable approach was found.

Electron impact can be assumed to be the major process contributing to hydrocarbon dissociation [1]. It is certainly the main process producing free radicals e.g. CH_3, CH_2, CH, OH and H, from hydrocarbon precursors. The elimination of a hydrogen atom from methane and an OH radical from methanol by electron impact are processes with large rate constants [1], but precise data for the conditions relevant to the used microwave plasma are not available. Donohue et al. showed that methanol is more readily dissociated than methane but their technique relates only to the formation of excited states, i.e., H* from CH_4 and OH* from methanol [173]. Nevertheless the C-O bond in methanol is weaker than the C-H bond in methane, and the methyl radical is a product in both cases. There is also a general trend for the electron impact dissociation rates to increase as the molecular weight of the hydrocarbon increases. Kim and co-workers compared the dissociation of water and methanol in rf plasmas and found complete methanol dissociation at the relatively low rf power of 300 W whereas water was only 20% dissociated at this power. These differences may be due to different dissociation energies [73].

Two basic experiments were done in a microwave discharge containing methanol and Argon only. The appearance of species should be dominated by the precursor methanol and its direct conversion counterparts. Therefore the concentrations of the methyl radical and formaldehyde versus the methanol admixture can be seen in Figure 8.41.

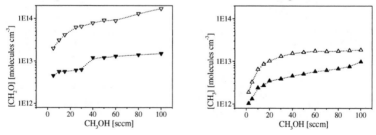

Figure 8.41   *Formaldehyde ($\triangledown$, $\blacktriangledown$) (fc,sc) and the methyl radical ($\triangle$, $\blacktriangle$)(fc,sc) in Ar-CH_3OH microwave plasmas with various amounts of methanol (P=1.5 kW, p=1.5 mbar).*

The values under sc are always lower than that under fc. This shows clearly that both molecules are intermediates. The amount of formaldehyde exceeds that of the methyl radical with about one order of magnitude. More important is the fact that the behaviour of both species is similar when increasing the amount of methanol. It seems that the formation of the molecules is related. The first contributing process is the electron impact dissociation of methanol into methyl radical and hydroxyl (OH) radical. Further reactions of OH radicals have fairly large rate constants, like the self recombination for water formation (K_{1000K} = 6.5 x 10^{-12} cm^{3}s^{-1}) [155]. Maybe also the reaction again with the precursor can appear. But with respect to the formation of formaldehyde the most likely process is the reaction with the methyl radical (CH$_3$ + OH → CH$_2$O + H$_2$: K_{1000K} = 1.3 x 10^{-11} cm^{3}s^{-1}) [161]. Thus immediately after the dissociation of methanol a fairly high content of formaldehyde can be observed. Its concentrations in H$_2$-Ar-CH$_3$OH, O$_2$-Ar-CH$_3$OH and Ar-CH$_3$OH plasmas can be seen in Table 8.9.

gas mixture	H$_2$/Ar/CH$_3$OH = 82/10.8/7.2	O$_2$/Ar/CH$_3$OH = 82/10.8/7.2	Ar/CH$_3$OH = 92.8/7.2
[CH$_2$O] [cm^{-3}]	3.4 x 10^{12}	6.1 x 10^{12}	1.2 x 10^{13}

Table 8.9   Formaldehyde concentrations in microwave plasmas (P=1.5 kW, p=1.5 mbar) using different source gas mixtures

Surprisingly, the highest formaldehyde concentration was found in Ar-CH$_3$OH plasmas, compare Table 8.9. This is probably due to the lack of atomic and molecular oxygen and hydrogen. Obviously, reactions with oxygen and hydrogen contribute significantly to the production and consumption of formaldehyde. As soon as higher amounts of hydrogen or oxygen are available the concentration of formaldehyde is decreased, in fact more decreased in the H$_2$-Ar-CH$_3$OH plasma.

The reaction scheme for the H$_2$-O$_2$-Ar-CH$_3$OH plasma deduced empirically from the experimental results with respect to the results in the methane-containing plasma can be seen in Figure 8.42.

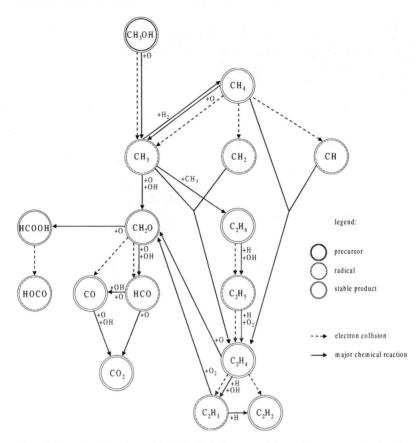

Figure 8.42   Reaction scheme of the H_2-Ar-O_2 discharge containing small percentages of methanol

As mentioned above the initial dissociation process of methanol is the electron collision. But also the reaction with oxygen seems to contribute additionally as shown by the increasing degree of dissociation with higher oxygen amounts in the source gas. The remarkable high methane concentration can be explained by the recombination of methyl radicals with H_2 ($K_{1000K} = 1.3 \times 10^{-12}$ cm^{3}s^{-1}). The methane concentration follows the behaviour of the H_2 partial pressure. For these plasmas the central position of the methyl radical was also confirmed. Its reactions lead to hydrocarbons and to formaldehyde. In case of increasing oxygen amounts in the source gas formic acid was also found. It is expected to be produced from formaldehyde due to the related increasing concentration behaviour when increasing the oxygen content. The HCO radical and carbon monoxide are produced by formaldehyde reactions. Thus formaldehyde is an important and very reactive intermediate.

Some hydrocarbon-containing reactions are expected to occur like in methane-containing discharges because of the remarkable content of methane and C-2 hydrocarbons measured in the methanol-containing discharge especially using source gas mixtures with less or without O_2. Similar exponential decays were found in the methanol-containing plasmas like in methane-containing plasmas, as shown in Table 6.5 and Table 6.6. The main behaviour of molecule concentration of carbon monoxide and carbondioxide as well as water is also comparable to methane-containing discharges, whereas the amounts of carbon monoxide and carbon dioxide are higher. The water concentration is almost insensitive to the hydrocarbon precursor. In methanol-containing discharges the maximum concentration of water at the 1:1 ratio of hydrogen and oxygen can also be observed. The exceptions are the conditions with either low hydrogen or oxygen content. Then the water concentration is also effected by the amount of methanol, compare chapter 6.4.4. The formaldehyde concentration is more than one order of magnitude higher in methanol-containing discharges.

This reaction scheme has yet to be verified by a detailed kinetic modelling. Nevertheless it is useful for a first intention about processes in a methanol-containing plasma.

9   Conclusion

The observation of the methyl radical and ten related stable species by TDLAS not only shows the versatility of the technique, but has enabled progress to be made in probing the species present and their concentrations under a wide range of oxidizing conditions. For the first time the absolute concentrations of a large group of molecules was measured in H_2-O_2-Ar microwave plasmas containing small percentages of hydrocarbon precursors either methane or methanol. Their degree of dissociation increased when using oxygen-containing source gas mixtures. The concentration of the methyl radical was found to be in the range from 10^{10} to 10^{12} cm^{-3}. The methyl radical and the stable products CH_4, C_2H_2, C_2H_4, C_2H_6 showed exponential decays with increasing oxygen amounts. For the first time methane was found as a plasma product in methanol-containing plasmas, as well as formic acid. Formaldehyde was found for the first time in such microwave plasmas containing methane or methanol. Further stable products are carbon monoxide and carbon dioxide. Under conditions with a lot oxygen the concentration of carbon dioxide exceeded that of carbon monoxide by more than one order of magnitude. The water concentration was measured to be nearly independent of the precursor having a maximum at about the 1:1 ratio of hydrogen to oxygen. In order to study the nature of the plasma chemistry all the measurements were done under both, flowing (fc) and static (sc) conditions.

The most significant differences of plasma chemical reactions between methane and methanol as hydrocarbon precursors are:

(i) the greater dissociation of methanol leading to higher concentrations of carbon dioxide and formaldehyde,

(ii) the formation of methane concentrations up to 10^{14} cm^{-3} in methanol plasmas in the absence of oxygen,

(iii) the gradual increase in formic acid concentration with oxygen in the methanol plasmas,

(iv) the absence of any detectable methanol or formic acid down to the detection limit of 10^{10} cm^{-3} in methane plasmas.

Mass balances were determined for both methane-containing and methanol-containing plasmas. It has been found that only at no or low oxygen amounts the hydrocarbons contribute to it, in fact at most in H_2-Ar-CH_4 plasmas. With increasing oxygen content the mass balance is dominated by carbon monoxide, carbon dioxide and water. Using moderate and high oxygen contents the sum of CO and CO_2 takes 50 % of the consumed carbon in the methane-containing discharge and about 80 % in the methanol-containing discharge.

Kinetic modelling has been carried out for H_2-O_2-Ar-CH_4 plasmas. An elaborate set of chemical reactions for the neutral species of the complex plasma was derived. The model calculations were performed with the set of chemical reactions available from the literature and these calculated dissociation rate coefficients. For the first time the comparison of calculated and measured species concentrations of the methyl radical and ten further stable molecules was possible.

Based on the comparison between measured and calculated concentrations the modelling was able:

(i) to achieve very good agreement for the H_2-Ar-CH_4 mixture and satisfactory agreement for the O_2-Ar-CH_4 mixture.

(ii) to determine a reaction scheme of H_2-Ar-CH_4 plasmas containing the most relevant neutral reactions by analysing the balanced production and consumption processes.

(iii) to derive for the first time a scheme for the main reactions in O_2-H_2-Ar-CH_4 plasmas.

(iv) to predict the concentrations of the further important radicals like e.g. OH, HO_2, HCO and CH_2 beside the molecules detected in the plasma so far.

Because of the promising results the modelling should be improved further especially for the plasma conditions containing oxygen.

Measurements of formaldehyde in accordance with the methyl radical were done in Ar-CH_3OH microwave plasmas. Both species were identified as intermediates. Results were used to get an idea for the initiating dissociation processes in methanol-containing plasmas. This approach and the experience from the methane-containing plasmas supported the establishment of a reaction scheme deduced from the experimental results containing the most important steps linking the plasma chemistry. This scheme represents a good basis for future studies concerning methanol-containing plasmas.

10   Future Work

Although this is the first time such a wide range of species concentrations have been monitored in H_2-Ar plasmas containing small amounts of methane or methanol with oxygen, it is clear that extension of the present studies to other intermediates is desirable. These include other free radicals the most important of which are OH, HCO, HO_2 and C_2H. The extension of TDLAS measurements to these species is feasible and in progress. To improve the modelling potential much more kinetic data on elementary reactions are required as well as quantitative measurements of the electron density and electron energy distribution functions.

Improvement of the optical set-up. With respect to the fairly low absorption coefficients of the molecule of interest or low radical concentrations in the plasma it will become necessary to use a long path cell, like it has been done in the majority of reported investigations. These cells allow multiple reflections inside the investigation volume for extending the absorption length. The sensitivity of the method can be increased over one to two orders of magnitude depending on the path length. For detection of radicals the plasma must be placed inside the reflection cell, thus the installation at the plasma vessel is necessary.

Using such a long path cell the predicted radical concentrations of OH, HO_2, CHO, C_2H_3, C_2H_5, C_2H and CH_2 obtained by the model calculations can be proved. Promising spectral positions can be seen in Table 4.1 in chapter 4.1.

Rapid scan technology. In order to exploit the capabilities of infrared tunable diode laser absorption spectroscopy for effective and reliable on-line plasma diagnostics and process control in research and industry, a compact and transportable tunable infrared multi-component acquisition system, "IRMA", has been developed. The IRMA system contains four independent laser stations which can be temporally multiplexed. A multi-path cell is included for exhaust gas detection. The special rapid scan software, developed at Aerodyne Research, Inc., is used [174,175].

The data acquisition method is an advanced form of sweep integration which is carried out by a software package. The software sweeps the laser frequency over the full infrared transition or group of transitions, then integrates the area under the transitions using non-linear least squares fitting to the known spectral line shapes and positions. Frequency modulation (FM) techniques are not used, because the clear connection between the direct absorption spectrum and the species' concentration is preferred.

There are several advantages to this sweep integration approach. First, absolute species concentrations are returned from the non-linear least squares fits so that external calibration is not required. The species concentrations may be determined from the absolute spectroscopic data available from the HITRAN or GEISA data bases or from user supplied data. Second, the line shape functions are known from theory and can be precisely calculated. Finally, this detailed understanding of the expected line shapes and positions allows the operator to easily monitor complex and overlapping spectral features using "fingerprint fitting". This is important because monitoring several transitions for one species can enhance sensitivity and is sometimes necessary - especially for larger molecules. In addition, fingerprint fitting allows one to monitor multiple species simultaneously, since overlapping lines can be used, and it even can allow the fitting of unknown lines which overlap the desired spectrum as a method of removing background absorption from unknown species.

The detector output voltage is sampled by an analog-to-digital converter using the data acquisition board. The individual spectra are automatically transferred to the computer's extended memory using direct memory access. The program divides the data in extended memory into individual sweeps and averages the sweeps to produce one resultant spectrum for each laser. The resultant spectra are analysed spectroscopically to determine the concentrations or column densities of any species which absorb in that spectral window. The column densities are displayed to the operator, saved to disk and analysed.

The multipass astigmatic Herriot cell with 36 m path length has been chosen, because it provides a minimum of cell volume for a given number of paths, and therefore optimized conditions for time-resolved measurements [176]. The used cell has a volume of 0.3 l, which allows, in principle, to reduce the residence time of exhaust gases in the cell to a few milliseconds, depending on the pumping system. In Figure 10.43 an example of sensitive detection of NH_3 is shown.

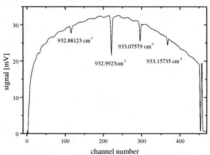

Figure 10.43   Example of a NH₃ absorption spectrum measured in the multipass Herriot cell, (N₂ + 20 ppm NH₃, p = 0.01 mbar, l = 36 m). [174]

An example spectrum measured in a H_2-CH_4 plasma excited in a GEC reactor [177,178] is shown in Figure 10.44. Acetylene and methane were measured by recording the change in absorption of individual rotational-vibrational lines at a frequency 1303 cm^{-1} (7.67 μm). The diode laser was scanned over a spectral region of 0.2 cm^{-1} by varying the laser current in order to obtain a baseline for the pair of spectral lines. The laser was turned off at the end of each scan to obtain the total laser intensity during the scan.

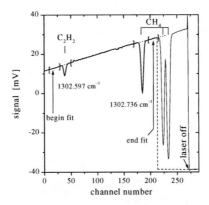

Figure 10.44   *Measured and fitted absorption spectrum of methane and acetylene in a H_2-CH_4*
plasma of a GEC reactor ($[H_2]$:$[CH_4]$ = 1:1, ϕ_{total} = 66 sccm, p = 100 Torr, P = 50 W).
[177,178]

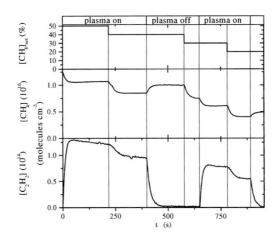

Figure 10.45   *Results of stream mode data acquisition of species densities with varying methane*
admixture using the absorption lines from Figure 10.44. [177,178]

Figure 10.45 shows, as an example, the result of a stream mode of data acquisition giving an example of online measurements useful especially for process control. In the graph one can see how the change of methane admixture to hydrogen and switching the plasma on/off in a GEC reactor influences the species densities of CH_4 and C_2H_2 (which were calculated using the lines shown in Figure 10.44) in the laser beam path.

An example of the burst data acquisition mode is shown in Figure 10.46. This method provides the highest possible time-resolution by transferring fifty measured spectra directly to the extended memory of the computer following a trigger pulse. The time-dependence of the dissociation of methane and the production of methyl radicals and C-2 hydrocarbons after switching on the plasma in a H_2-CH_4 microwave surface wave discharge has been analysed. The concentration of the C-2 products increases with time, while the density of the methyl radicals remains constant within the first seconds of the discharge. Even at a time resolution of 50 ms the CH_3 density was found to be nearly constant in the discharge tube [174,179].

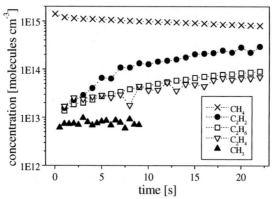

Figure 10.46   Results of burst mode data acquisition of species densities with time in a microwave surface wave discharge, (H_2+ 10% CH_4, p= 1 Torr, P= 600 W). [179]

The diode lasers have also been developed further. Different material compositions have been used. Recently, first results of so-called Quantum Cascade Lasers (QCL) were published [180]. These very promising radiation sources are characterised by stronger emission intensities, about hundret times stronger than that of lead salt diode lasers, but on the other hand their spectroscopic properties as e.g. tunability are still poor.

11 References

[1] L.E. Kline, W.D. Partlow and W.E. Bies, *J. Appl. Phys.* **65**, 70 (1989).

[2] D.J. Dagel, C.M. Mallouris and J.R. Doyle, *J. Appl. Phys.* **79**, 8735 (1996).

[3] L. Okeke and H. Störi, *Plasma Chem. Plasma Process.* **11**, 489 (1991).

[4] J.L. Jauberteau, L. Thomas, J. Aubreton, I. Jauberteau and A. Catherinot, *Plasma Chemistry and Plasma Processing* **18**, 137 (1998).

[5] C. Benndorf, P. Joeris and R. Kröger, *Pure & Applied Chemistry* **66**, 1195 (1994).

[6] H. Sugai, H. Kojima, A. Ishida and H. Toyoda, *Appl. Phys. Lett.* **56**, 2616 (1990).

[7] H. Sugai and H. Toyoda, *J. Vac. Sci. Technol. A* **8**, 1193 (1992).

[8] S. Ando, M. Shinohara and K. Takayama, *Vacuum* **49**, 113 (1998).

[9] P. Pecher and W. Jacob, *Appl. Phys. Lett.* **73**, 31 (1998).

[10] M. Schmidt, R. Foest and R. Basner, *J. Phys. IV France* **8**, 231 (1998).

[11] H. Budzikiewicz, *Massenspektrometrie: Eine Einführung*, VCH, Weinheim, (1992).

[12] M.H. Kim and J.Y. Lee, *Journal of Material Science* **26**, 4787 (1991).

[13] S. Harris and A.M. Weiner, *J. Appl. Phys.* **67**, 6520 (1990).

[14] E. Leibnitz und H.G. Struppe, *Handbuch der Gaschromatographie*, Leipzig, (1970).

W. Rödel und G. Wölm, *Grundlagen der Gaschromatographie*, DVW Berlin, (1982).

[15] J. Röpcke, M. Käning, B.P. Lavrov and P.B. Davies, in *Lectures in Plasma Physics and Plasma Technology*, S. Pfau, M. Schmidt, R. Hippler and K.H. Schönbach Eds., Wiley-Verlag Chemie, Weinheim, 1999, in press.

[16] C. Benndorf, P. Joeris and R. Kröger, *Pure & Appl. Chem.* **66**, 1195 (1994).

[17] J. Röpcke and A. Ohl, *Contrib. Plasma Phys.* **34**, 575 (1994).

[18] W. Demtröder, *Laserspektroskopie: Grundlagen und Techniken*, Springer-Verlag Heidelberg Berlin, (1991).

[19] W. Neumann, in *Ergebnisse der Plasmaphysik und der Gaselektronik*, Band 1, Eds. R. Rompe und M. Steenbeck, Akademie-Verlag Berlin, (1967).

[20] R. De Avillez Pereira, D.L. Baulch, M.J. Pilling, S.H. Robertson, G. Zeng, *J. Phys. Chem. A* **101**, 9681 (1997).

[21] S.M. Hooker and C.E. Webb, Prog. Quant. Electron. 18, 227 (1994).

[22] U. Czarnetzki, K. Miyazaki, T. Kajiwara, K. Muraoka, M. Maeda and H.F. Döbele, *Appl. Opt.* **11**, 2155 (1994).

[23] T. Doerk, J. Ehlbeck, P. Jauernik, J. Staúco, J. Uhlenbusch and T. Wottka, *J. Phys D: Appl. Phys.* **26**, 1015 (1993).

[24] V. Kornas, V. Schulz-von der Gathen, T. Bornemann and H.F. Döbele, *Plasma Chem. Plasma Process.* **11**, 171 (1991).

[25] V. Kornas, A. Roth, H.F. Döbele and G. Pross, *Plasma Chem. Plasma Process.* **15**, 71 (1995).

[26] S. Hädrich, B. Pfelzer and J. Uhlenbusch, *Plasma Chem. Plasma Process.* **19**, 91 (1999).

[27] T. Doerk, Dissertation, Heinrich-Heine-Universität Düsseldorf, *Wissenschaftliche Schriftenreihe* **46**, (1996).

[28] H. Conrads, *Zeitschrift für Physik* **200**, 444 (1967).

[29] A. O'Keefe and D.A.G. Deacon, *Rev. Sci. Instrum.* **59**, 2544 (1988).

[30] A. Campargue, D. Romanini, N. Sadeghi, *J. Phys. B: Appl. Phys.* **31**, 1168 (1998).

[31] M. Kotterer, J. Conceicao, J. P. Maier, *Chem. Phys. Lett.* **259**, 233 (1996).

[32] G. Meijer, M.G.H. Boogaarts, R.T. Jongma, D.H. Parker and A.M. Wodtke, *Chem. Phys. Lett.* **217**, 112 (1994).

[33] P. Zalicki, Y. Ma, R.N. Zare, E.H. Wahl, J.R. Dadamio, T.G. Owano and C.H. Kruger, *Chem. Phys. Lett.* **234**, 269 (1995).

[34] R. Engeln, G. Berden, R. Peeters and G. Meijer, Centre de Recherches en Physique des Plasmas, EPFL, *Lausanne Report LRP* **629/99** „Workshop on Frontiers in Low Temperature Plasma Diagnostics III", Switzerland, 15-19 February 1999, Book of Papers, 39.

[35] R. Engeln, E. van den Berg, G. Meijer, L. Lin, G.M.H. Knippels and A.F.G. van der Meer, *Chem. Phys. Lett.* **269**, 293 (1997).

[36] G.M.W. Kroesen, J.H.W.G. den Boer, L. Boufendi, F. Vivet, K, Khouli, A. Bouchoule and F.J. de Hoog, *J. Vac. Sci. Technol. A* **14**, 546 (1996).

[37] P.R. Griffiths and J.A. de Haseth, Fourier Transformed Infrared Spectroscopy, Eds. P.J. Elving and J.D. Winefordner, John Wiley & Sons, New York (1986).

[38] J. Wormhoudt, *J. Vac. Sci. Technol. A* **8**, 1722 (1990).

[39] P.B. Davies and P.M. Martineau, *Appl. Phys. Lett.* **57**, 237 (1990).

[40] M. Haverlag, E. Stoffels, W.W. Stoffels, G.M.W. Kroesen and F.J. de Hoog, *J. Vac. Sci. Technol. A* **14**, 380 (1996).

[41] P.B. Davies and P.M. Martineau, *Adv. Mater.* **4**, 729 (1992).

[42] S. Naito, N. Ito, T. Hattori and T. Goto, *Jpn. J. Appl. Phys.* **34**, 302 (1995).

[43] M. Haverlag, E. Stoffels, W.W. Stoffels, G.M.W. Kroesen and F.J. de Hoog, *J. Vac. Sci. Technol. A* **12**, 3102 (1994).

[44] J. Röpcke, L. Mechold, M. Käning, W.Y. Fan and P.B. Davies, *Plasma Chem. Plasma Process.* **19**, 395 (1999).

[45] M. Schmidt, R. Foest and R. Basner, *J. Phys. IV France* **8**, 231 (1998).

[46] P. Pecher, Dissertation, Max-Planck-Institut für Plasmaphysik Garching, (1998).

[47] F.G. Celii and J.E. Butler, *J. Appl. Phys.* **71**, 2877 (1992).

[48] G. Herzberg and J. Shoosmith, *Can. J. Phys.* **34**, 523 (1956).

[49] G. Herzberg, *Proc. Roy. Soc. A* **262**, 291 (1961).

[50] L.Y. Tan, A.M. Winer and G.C. Pimentel, *J. Chem. Phys.* **57**, 4028 (1972).

[51] C. Yamada, E. Hirota and K. Kawaguchi, *J. Chem. Phys.* **75**, 5256 (1981).

[52] T. Amano, P.F. Bernath, C. Yamada, Y. Endo, and E. Hirota, *J. Chem Phys.* **77**, 5284 (1982).

[53] C. Yamada and E. Hirota, *J. Chem. Phys.* **78**, 669 (1983).

[54] G.A. Laguna and S.L. Baughcum, *Chem. Phys. Lett.* **88**, 568 (1982).

[55] G.A. Bethardy and R.G. Macdonald, *J. Chem. Phys.* **103**, 2863 (1995).

[56] G.A. Bethardy, F.J.Northrup and R.G. Macdonald, *J. Chem. Phys.* **105**, 5433 (1996).

[57] D.F. Davidson, M.D. Di Rosa, E.J. Chang and R.K. Hanson, *J. Quant. Spectrosc. Radiat. Transfer* **53**, 581 (1995).

[58] D.H. Fairbrother, K.A. Briggman, K.A. Dickens, P.C. Stair and E. Weitz, *Rev. Sci. Instrum.* **68**, 2031 (1997).

[59] G.N. Robinson, M.S. Zahniser, A. Freedman and D.D. Nelson Jr., *J. Mol. Spectr.* **176**, 337 (1996).

[60] T. Takayanagi, *J. Chem. Phys.* **104**, 2237 (1996).

[61] K. Tachibana, M. Nishida, H. Harima and Y. Urano, *J. Phys. D: Appl. Phys.* **17**, 1727 (1984).

[62] F.G. Celii, P.E. Pehrsson, H.-T. Wang and J.E. Butler, *Appl. Phys. Lett.* **52**, 2043 (1988).

[63] K.L. Menningen, M.A. Childs, H. Toyoda, Y. Ueda, L.W. Anderson and J.E. Lawler, *Diamond & Related Materials* **3**, 422 (1994).

[64] M.A. Childs, K.L. Menningen, P. Chevako, N.W. Spellmeyer, L.W. Anderson and J.E. Lawler, *Phys. Lett. A* **171**, 87 (1992).

[65] J. Wormhoudt and K.E. McCurdy, *Chem. Phys. Lett.* **156**, 47 (1989).

[66] P.B. Davies and P.M. Martineau, *J. Appl. Phys.* **71**, 6125 (1992).

[67] S. Naito, H. Nomura and T. Goto, *Rev. Laser Eng.* **20**, 746 (1992).

[68] S. Naito, N. Ito, T. Hattori and T. Goto, *Jpn. J. Appl. Phys.* **33**, 5967 (1994).

[69] S. Naito, M. Ikeda, N. Ito, T. Hattori and T. Goto, *Jpn. J. Appl. Phys.* **32**, 5721 (1993).

[70] M. Ikeda, N. Ito, M. Hiramatsu, M. Hori and T. Goto, *J. Appl. Phys.* **82**, 4055 (1997).

[71] M. Ikeda, K. Aiso, M. Hori and T. Goto, *Jpn. J. Appl. Phys.* **34**, 3273 (1995).

[72] M. Ikeda, M. Hori, T. Goto, M. Inayoshi, K. Yamada, M. Hiramatsu and M. Nawata, *Jpn. J. Appl. Phys.* **34**, 2484 (1995).

[73] S. Kim, D.P Billesbach and R. Dillon, *J. Vac. Sci. Technol. A* **15**, 2247 (1997).

[74] M. Moisan and J. Pelletier, Microwave excited Plasmas, *Plasma Technology* **4**, Elsevier Science Publishers B.V., Netherlands, (1992).

[75] M. Füner, C. Wild and P. Koidl, *Surface and Coatings Technology* **116-119**, 853 (1999).

[76] M.R. Wertheimer and M. Moisan, *J. Vac. Sci. Technol. A* **3**, 2643 (1985).

[77] R. Winkler in *Microwave Discharges: Fundamentals and Applications*, Edited by C.M. Ferreira and M. Moisan, Chapter: Collision Dominated Electron Kinetics in Low and High Frequency Fields, 339 (1993).

[78] A. Ohl in *Microwave Discharges: Fundamentals and Applications*, Edited by C.M. Ferreira and M. Moisan, Chapter: Large Area Planar Microwave Plasmas, 205 (1993).

[79] C. Rint, Herausgeber, Handbuch für Hochfrequenz- und Elektro-Techniker, Band 2, Kapitel 2, Hütig und Pflaum Verlag, 612 (1978).

[80] M. Kummer, Grundlagen der Mikrowellentechnik, Kapitel Wellenleiter, VEB Verlag Technik, 76 (1986).

[81] M. Käning, Dissertation, INP Greifswald an der Ernst-Moritz-Arndt-Universität Greifswald, (1997).

[82] A. Ohl, *J. Phys. IV France* **8**, Pr7-82 (1998).

[83] J. Röpcke and A. Ohl, *Contrib. Plasma Phys.* **31**, 669 (1991).

[84] A. Ohl, *Pure & Appl. Chem.* **66**, 1397 (1994).

[85] J. Röpcke, *Contrib. Plasma Phys.* **34**, 575 (1994).

[86] R.T. Menzies, J.S. Margolis, E.D. Hinkley and R.A. Toth, *Appl. Opt.* **16**, 523 (1977).

[87] H. Jost and M. Loewenstein, 5th Int. Symp. on Gas Analysis by Tunable Diode Lasers, Februar 1998, *VDI Berichte* 1366, 93 (1998).

[88] F.-J. Lübken, F. Dingler and H.v. Lucke, 5th Int. Symp. on Gas Analysis by Tunable Diode Lasers, Februar 1998, *VDI Berichte* 1366, 101 (1998).

[89] P. Werle, 5th Int. Symp. on Gas Analysis by Tunable Diode Lasers, Febr. 1998, *VDI Berichte* 1366, 1 (1998).

[90] J. Reid and D.T. Cassidy, *Appl. Opt.* **21**, 3961 (1982).

[91] A.D. Richards, B.E. Thompson, K.D. Allan and H.H. Sawin, *J. Appl. Phys.* **62**, 792 (1987).

[92] P.B. Davies and D.K. Russell, *Chem. Phys. Lett.* **67**, 440 (1979).

[93] G.M.W. Kroesen and F.J. de Hoog, *Appl. Phys. A* **56**, 479 (1993).

[94] K. Takahashi, M. Hori and T. Goto, *J. Vac. Sci. Technol. A* **14**, 2004 (1996).

[95] K. Takahashi, M. Hori and T. Goto, *J. Vac. Sci. Technol. A* **14**, 2011 (1996).

[96] P.F. Bernath, *Annu. Rev. Phys. Chem.* **41**, 91 (1990).

[97] E. Hirota, *High-Resolution Spectroscopy of Transient Molecules*, Springer-Verlag, Berlin Heidelberg, (1985).

[98] W.Y. Fan, J. Röpcke and P.B. Davies, *J. Vac. Sci. Technol. A* **14**, 2970 (1996).

[99] W.Y. Fan, P.F. Knewstubb, M. Käning, L. Mechold, J. Röpcke and P.B. Davies, *J. Phys. Chem. A* **103**, 4118 (1999).

[100] W. Fan, Dissertation Thesis, University of Cambridge, Department of Chemistry, (1997).

[101] C.S. Gudeman, M.H. Begemann, J. Pfaff and R.J. Saykally, *Phys. Rev. Lett.* **50**, 727 (1983).

[102] Z. Liu and P.B. Davies, *Phys. Rev. Lett.* **76**, 596 (1996).

Z. Liu and P.B. Davies, *J. Chem. Phys.* **105**, 3443 (1996).

[103] D.M. Smith, P.M. Martineau and P.B. Davies, *J. Chem. Phys.* **96**, 1741 (1992).

P.B. Davies and D.M. Smith, *J. Chem. Phys.* **100**, 6166 (1994).

[104] Z. Liu, Dissertation Thesis, University of Cambridge, Department of Chemistry, (1997).

[105] K. Matsumura, H. Kanamori, K. Kawaguchi and E. Hirota, *J. Chem. Phys.* **89**, 3491 (1988).

[106] K.H. Hinkle, J.J. Keady and P.F. Bernath, *Science* **241**, 1319 (1988).

[107] N. Moazzen-Ahmadi, A.R.W. McKellar and T. Amano, *Chem. Phys. Lett.* **157**, 1 (1989).

[108] N. Moazzen-Ahmadi, A.R.W. McKellar and T. Amano, *J. Chem. Phys.* **91**, 2140 (1989).

[109] J.R. Heath, A.L. Cooksy, M.H.W. Gruebele, C.A. Schmuttenmaer and C.A. Saykally, *Science* **244**, 564 (1989).

[110] P.F. Bernath, K.H. Hinkle and J.J. Keady, *Science* **244**, 562 (1989).

[111] A.R.W. McKellar, C. Yamada and E. Hirota, *J. Chem. Phys.* **79**, 1220 (1983).

[112] M. Marshall and A.R.W. McKellar, *J. Chem. Phys.* **85**, 3716 (1986).

[113] H. Kanamori, K. Seki and E. Hirota, *J. Chem. Phys.* **87**, 73 (1987).

[114] H. Kanamori and E. Hirota, *J. Chem. Phys.* **89**, 3962 (1988).

[115] K. Kawaguchi, T. Amano and E. Hirota, *J. Mol. Spectrosc.* **131**, 58 (1988).

[116] H. Kanamori, Y. Endo and E. Hirota, *J. Chem. Phys.* **92**, 197 (1990).

[117] J.-H. Wang, H.-C. Chang and Y.-T. Chen, *J. Chem. Phys.* **206**, 43 (1996).

[118] T.J. Sears and J.M. Frye, *44th Symp. Mol. Spectrosc.*, Columbus, Ohio, Abstr. WF5 (1989).

[119] Y. Hirahara, A. Masuda and K. Kawaguchi, *J. Chem. Phys.* **95**, 3975 (1991).

[120] C.B. Dane, D. R. Lander, R.F. Curl, F.K. Tittel, Y. Guo, I.F. Ochsner and C.B. Moore, *J. Chem. Phys.* **88**, 2121 (1988).

[121] B.M. Landsberg, A.J. Merer and T. Oka, *J. Mol. Spectrosc.* **67**, 459 (1977).

[122] J.W.C. Johns, A.R.W. McKellar and M. Riggin, *J. Chem. Phys.* **67**, 2427 (1977).

[123] J.M. Brown, J. Buttenshaw, A. Carrington, K. Dumper and C.R. Parent, *J. Mol. Spectrosc.* **79**, 47 (1980).

[124] J.M. Brown, K. Dumper and R.S. Lowe, *J. Mol. Spectrosc.* **97**, 441 (1983).

[125] A.R.W. McKellar, J.B. Burkholder, J.J. Orlando and C.J. Howard, *J. Mol. Spectrosc.* **130**, 445 (1988).

[126] C. Yamada, H. Kanamori, H. Horiguchi, S. Tsuchiya and E. Hirota, *J. Chem. Phys.* **84**, 2573 (1986).

[127] K. Nagai, Y. Endo and E. Hirota, *J. Mol. Spectrosc.* **89**, 520 (1981).

[128] M.S. Zahniser, K.E. McCurdy and A.C. Stanton, *J. Phys. Chem.* **93**, 1065 (1989).

[129] J.W.C. Johns, A.R.W. McKellar and M. Riggin, *J. Chem. Phys.* **68**, 3957 (1978).

[130] P.B. Davies and D.K. Russell, *J. Mol. Struct.* **60**, 201 (1980).

[131] M.S. Zahniser and A.C. Stanton, *J. Chem. Phys.* **80**, 4951 (1984).

[132] D.D. Nelson, Jr. and M.S. Zahniser, *J. Mol. Spectrosc.* **150**, 527 (1991).

[133] T.W. Ducas, L.D. Geoffrion, R.M Osgood Jr. and A. Javan, *Appl. Phys. Lett.* **21**, 42 (1972).

[134] B. Lemoine, M. Bogey, J.L. Destombes, *Chem. Phys. Lett.* **117**, 532 (1985).

[135] Laser Components Booklet, *IR Laser Diodes 3-25 μm*, Laser Components GmbH, 2 (1998).

[136] U. Schießl and M. Hodges, private communication, (1999).

[137] U. Schießl and H.E. Wagner, 5th Int. Symp. on Gas Analysis by Tunable Diode Lasers, Februar 1998, *VDI Berichte* 1366, 251 (1998).

[138] G. Guelachvili and K.N. Rao, *Handbook of Infrared Standards*, Academic Press, Orlando, 1986.

[139] L.S. Rothman et al., *J. Quant. Spectr. Radiat. Transfer* **48**, 469 (1992).

[140] N. Husson et al., *J. Quant. Spectr. Radiat. Transfer* **52**, 425 (1994).

[141] N. Jacquinet-Husson et al., *J. Quant. Spectr. Radiat. Transfer* **62**, 205 (1999).

[142] D. Rehle, Diplomarbeit, INP Greifswald an der Ernst-Moritz-Arndt-Universität Greifswald, (1995).

[143] P. Werle, *Spectrochimica Acta Part A* **52**, 805 (1996).

[144] C. Busch, Dissertation, Ruhr-Universität-Bochum, (1999).

[145] A.N. Dharamsi, *J. Phys. D: Appl. Phys.* **25**, 540 (1996).

[146] A. Ohl, *Microwave Discharges: Fundamentals and Application*, C.M. Ferreira (Ed.), Plenum Press, New York, 205 (1993).

[147] C. Brechignac, J.W. Johns, A.R.W. McKellar and M. Wong, *J. Mol. Spectrosc.* **96**, 353 (1982).

[148] D.G. Murcray, F.J. Murcray, F.S. Bonomo, A. Goldman and R.D. Blatherwick, „*High Resolution Infrared Laboratory Spectra*", University of Denver, Denver, Colorado (1984).

[149] J. Röpcke, M. Käning and B.P. Lavrov, *J. Phys. IV France* **8**, Pr7-207 (1998).

[150] S.A. Astashkevich, M. Käning, E. Käning, N.V. Kokina, B.P. Lavrov, A. Ohl and J. Röpcke, *J. Quant. Spectr. Radiat. Transfer* **56**, 725 (1996).

[151] J. Röpcke, B.P. Lavrov and A. Ohl, Frontiers in Low Temperature Plasma Diagnostics II, Bad Honnef, *Book of Papers*, 201 (1997).

[152] B.P. Lavrov, M. Käning, V.L. Ovtchinnikov and J. Röpcke, Frontiers in Low Temperature Plasma Diagnostics II, Bad Honnef, *Book of Papers*, 169 (1997).

[153] J. Röpcke, L. Mechold, M. Käning, W.Y. Fan and P.B. Davies, 5th Int. Symp. on Gas Analysis by Tunable Diode Lasers, Februar 1998, *VDI Berichte* 1366, 155 (1998).

[154] H.F. Winters, *J. Chem. Phys.* **63**, 3462 (1975).

[155] D.L. Baulch et al., *J. Phys. Chem. Ref. Data* **21**, 411 (1992).

[156] P.K. Bachmann, D. Leers and H. Lydtin, *Diamond & Related Materials* **1**, 1 (1991).

[157] W.L. Gardner, *J. Vac. Sci. Technol. A* **14**, 1938 (1996).

[158] C.G. Schwärzler, O. Schnabl, J. Laimer and H. Störi, *Plasma Chem. Plasma Process.* **16**, 173 (1996).

[159] W. Möller, *Appl. Phys. A* **56**, 527 (1993).

[160] T. Lang, J. Laimer and H. Störi, *Diam. Rel. Mater.* **3**, 470 (1994).

[161] M. Frenklach and H. Wang, *Phys. Rev. B* **43**, 1520 (1991).

[162] H. Pauser, C.G. Schwärzler, J. Laimer and H. Störi, *Plasma Chem. Plasma Process.* **17**, 107 (1997).

[163] L.S. Zelson, D.F. Davidson and R.K. Hanson, *J. Quant. Spectrosc. Radiat. Transfer* **52**, 31 (1994).

[164] W.L. Hsu, J. Appl. Phys. 72, 3102 (1992).

[165] L. Mechold, J. Röpcke, M. Käning, D. Loffhagen and P.B. Davies, Centre de Recherches en Physique des Plasmas, EPFL, *Lausanne Report LRP* **629/99** „Workshop on Frontiers in Low Temperature Plasma Diagnostics III", Switzerland, 15-19 February 1999, Book of Papers, 155.

[166] S. J. Buckman and A. V. Phelps, *JILA Information Center Report* **27**, University of Colorado, Boulder, Colorado (1985).

[167] M. Hayashi, Electron Collision Cross Sections, in: *Plasma Material Science Handbook*, Japan Society for the Promotion of Science (Ed.), Ohmsha, Ltd, Tokyo, 748 (1992).

[168] D. K. Davies, L. E. Kline and W. E. Bies, *J. Appl. Phys.* **65**, 3311 (1989).

[169] A. Levchenko and G. Alexeev, KINEL Manual, Moscow (1994).

[170] G.Y. Alekseev, V.A. Bityrin, A.N. Bocharov, A.V. Levchenko, A. Ohl and M. Schmidt, *Proc. 12th Int. Conf. Gas Discharges and Their Applications*, Greifswald, 407 (1997).

[171] W.C. Gear, *Numerical Initial Value Problems in Ordinary differential equations*, Englewood Cliffs, NL: Prentice-Hall, (1971).

[172] FACSIMILE manual, AEA Technology Dorchester, (1999).

[173] D. E. Donohue, J. A. Schiavone, and R. S. Freund, *J. Chem. Phys.* **67**, 769 (1977).

[174] J. Röpcke, L. Mechold, M. Käning, J. Anders, F.G. Wienhold, D. Nelson and M. Zahniser, *Rev. Sci. Instum.*, in press (2000).

[175] J. Röpcke, M. Käning, J. Anders, F.G. Wienhold, D. Nelson and M. Zahniser, Centre de Recherches en Physique des Plasmas, EPFL, *Lausanne Report LRP* **629/99** „Workshop on Frontiers in Low Temperature Plasma Diagnostics III", Switzerland, 15-19 February 1999, Book of Papers, 113.

[176] J.B. McManus, P. L Kebabian and M. S. Zahniser, *Applied Optics* **34**, 3336 (1995).

[177] M. Käning, J. Röpcke, C. Lukas, T. Kawetzki, V. Schulz-von der Gathen, and H. F. Döbele, Centre de Recherches en Physique des Plasmas, EPFL, *Lausanne Report LRP* **629/99** „Workshop on Frontiers in Low Temperature Plasma Diagnostics III", Switzerland, 15-19 February 1999, Book of Papers, 257.

[178] C. Lukas, T. Kawetzki, V. Schulz-von der Gathen, H. F. Döbele, M. Käning and J. Röpcke, Centre de Recherches en Physique des Plasmas, EPFL, *Lausanne Report LRP* **629/99** „Workshop on Frontiers in Low Temperature Plasma Diagnostics III", Switzerland, 15-19 February 1999, Book of Papers, 261.

[179] L. Mechold, J. Röpcke, X. Duten, A. Rousseau, Frühjahrstagung Plasmaphysik der DPG, Bonn, *Verhandlungen der DPG* **5/2000**, 1024 (2000).

[180] Proceedings of the *Workshop of Quantum Cascade Lasers 03/2000*, IPM Freiburg, (2000).

Glossary

Used acronyms:

AS	absorption spectroscopy
CARS	coherent anti-Raman scattering
CRDS	cavity ring down spectroscopy
CVD	chemical vapor deposition
dc	direct current
DFWM	degenerate four wave mixing
ECR	electron cyclotron resonance
EEDF	electron energy distribution function
fc	flowing conditions
FIR	far infrared
FM	frequency modulation
FSR	free spectral range of the Fabry-Perot etalon
FTIR	Fourier transformed infrared
GC	gas chromatography
He system	TDL system with an included closed cycle Helium refridgerator for cooling of the diodes
IR	Infrared
IRMA	infrared multi-component acquisition system
LIF	laser induced flourescence
LN_2 system	TDL system with an included LN_2 evaporator for cooling of the diodes
MS	mass spectrometry
NIR	near infrared
OES	optical emission spectroscopy
PIMS	photon ionisation mass spectrometry
RECARS	resonance-enhanced coherent anti-Raman scattering
rf	radio frequency
sc	static conditions
TDL	tunable diode laser

TDLAS	tunable diode laser absorption spectroscopy
TIMS	threshold ionisation mass spectrometry
UV	ultraviolet
VIS	visible
VUV	vacuum ultraviolet, also called extreme ultraviolet

Used symbols:

I_v	light intensity at the frequency v
l	path length
$\varepsilon(v)$	emission coefficient at the frequency v
$\kappa(v)$	absorption coefficient at the frequency v
N_i	species population densities of the energy levels i
$\sigma_i(v)$	cross section for light absorption of the energy level i at the frequency v
v	frequency of the radiation
B_{ik}	Einstein coefficient for the transition between energy levels i and k
m_e	mass of the electron
e	elementary charge
$\vec{E}$	vector of the electric field strength
v_c	average electron-neutral collision frequency for momentum transfer
$\vec{v}_T$	translational velocity
$\vec{v}_{Tc}$	complex amplitude of the translational velocity
$\vec{J}$	current density
n_e	electron number density
σ	permittivity
ε	permeability
ε_0	permeability of the vacuum
ω_p	plasma frequency
$\bar{u}$	averaged picked up kinetic energy per period
$\hat{E}$	amplitude of the electric field
Θ_A	mean power transfer per electron
E/N	reduced field strength
$f(v-v_0)$	line shape function of a spectral line

$f_D(v-v_0)$	line shape function of a spectral line of the Doppler broadening usually described by the Gaussian function
$f_L(v-v_0)$	line shape function of a spectral line of the pressure broadening usually described by the Lorentzian function
Δv_D	half width at the half maximum amplitude of the Gaussian function (Doppler width)
Δv_L	half width at the half maximum amplitude of the Lorentzian function (Lorentz width)
Δv_{L0}	half width at the half maximum amplitude of the Lorentzian function (Lorentz width) at standard conditions (p_0, T_0)
c	vacuum light velocity
k	Boltzmann constant
N_A	Avogadro constant
v_0	spectral position of the absorption line
p	pressure
T	Temperature
$n, [species]$	species concentration, density
I_0	light intensity entering the destination volume
I	transmitted light intensity
S	line strength
S/N	Signal to Noise ratio
U_{ds}	detector voltage
U_m	modulated voltage as function of
A_m	modulation amplitude
λ	wavelength
d_g	gap distance
γ_i, γ'_i	stoichiometric coefficients
A_i, A'_i	reagents, products
C_k	one species
K_j	reaction rate coefficient of the j-th reaction
K_0	preexponential factor in the Arrhenius equation
E_A	activation energy
E_{ac}	field strength of the oscillating field
E_0	amplitude of the actual ac field

Dankeschön!

An dieser Stelle möchte ich mich bei allen bedanken, die Anteil am Gelingen dieser Arbeit hatten. Natürlich hätte ich jetzt viele Namen zu nennen, und gerade deswegen fällt es mir schwer mit einzelnen Namen zu beginnen.

Mein besonderer Dank gilt meinen Mentoren Prof. Dr. J. P. F. Conrads und Dr. J. Röpcke für das interessante Thema, das in mich gesetzte Vertrauen und die sehr gute Zusammenarbeit. Hier fand ich immer Bereitschaft zu wissenschaftlichen Diskussionen und Hilfe in kritischen Situationen. Auch an Dr. Paul Davies, University of Cambridge, geht ein großes Dankeschön für seine hilfreiche Unterstützung. Thanks a lot Dr. Davies for your support. Für die vielfältige Zusammenarbeit im experimentellen Bereich, die mir eine reibungslose Durchführung der Messungen gestattete, danke ich unserem Techniker D. Gött. Herzlichen Dank auch an Dr. D. Loffhagen und Dr. M. Schmidt für die interessanten wissenschaftlichen Diskussionen.

Darüber hinaus danke ich allen Kollegen, den Mitarbeitern des Instituts für Niedertemperatur-Plasmaphysik e.V., für ihre Zusammenarbeit. Das engagierte Zusammenspiel aller Abteilungen im Hause erzeute eine besondere Arbeitsatmosphäre und bot somit hervorragende Voraussetzungen für die Promotion.

Außerdem danke ich meiner Familie und allen Freunden, den einen für die tägliche Unterstützung vor Ort und den anderen für die freundliche und regelmäßige Erinnerung an meine eigenen Zielsetzungen. Bei ihnen fand ich immer wieder den nötigen Halt, aber auch Zerstreuung in schwierigen Situationen.

Hiermit erkläre ich, daß diese Arbeit bisher von mir weder an der Mathematisch-Naturwissenschaftlichen Fakultät der Ernst-Moritz-Arndt-Universität noch einer anderen wissenschaftlichen Einrichtung zum Zwecke der Promotion eingereicht wurde.

Ferner erkläre ich, daß ich diese Arbeit selbständig verfaßt und keine anderen als die darin angegebenen Hilfsmittel benutzt habe.

Tabellarischer Lebenslauf

Persönliches:

Name:	Mechold
Vorname:	Lars
geboren am:	26.04.1970
Geburtsort:	Pößneck (Thüringen)
Wohnort:	Greifswald
Adresse (dienstlich):	Institut für Niedertemperatur-Plasmaphysik e.V. (INP)
	F.-L.-Jahn-Str. 17, 17489 Greifswald
	Tel.: 03834/554429, Fax: 03834/554301
	e-mail: mechold@inp-greifswald.de
Adresse (privat):	Steinstraße 47, 17489 Greifswald
	Tel.: 03834/502708

Schulbildung:

1976 - 1986	Allgemeine Polytechnische Oberschule
1986	Abschluß der Allgemeinbildenden Oberschule in Staßfurt
1986 - 1988	Erweiterte Oberschule in Staßfurt
1988	Abitur mit dem Prädikat „Auszeichnung"
1988 - 1989	Wehrdienst

Hochschulausbildung:

1989 - 1995	Studium der Physik an der Universität Greifswald
Januar 1995	Abschluß des Studiums mit dem Diplom (Pradikat „Gut")
seit Februar 1995	Mitarbeiter am INP
seit November 1996	Doktorand mit dem vorliegenden Thema

List of Publications

D. Mönke, H. Scheibner, L. Mechold, W. Becker and F. Schrade, „Density Distribution of Plasma Species in a Hollow Cathode Arc Discharge for Coating Applications", *Contib. Plasma Phys.* **36**, 631-643 (1996).

J. Röpcke, L. Mechold, M. Käning, W.Y. Fan and P.B. Davies, „Diagnostics of Molecular Microwave Plasmas by Tunable IR Diode Laser Spectroscopy", 5th Int. Symp. on Gas Analysis by Tunable Diode Lasers, *VDI Berichte* **1366**, 155-164 (1998).

J. Röpcke, L. Mechold, M. Käning, W.Y. Fan and P.B. Davies, „Tunable Diode Laser Diagnostic Studies of H_2-Ar-O_2 Microwave Plasmas Containing Methane or Methanol", *Plasma Chem. and Plasma Process.* **19**, 395-419 (1999).

W.Y. Fan, P.F. Knewstubb, M. Käning, L. Mechold, J. Röpcke and P.B. Davies, „A Diode Laser and Modelling Study of Mixed (CH_4-H_2-O_2) AC Plasmas", *J. Phys. Chem. A* **103**, 4118-4128 (1999).

J. Röpcke, L. Mechold, M. Käning, J. Anders, F.G. Wienhold, D. Nelson, and M. Zahniser, „IRMA: a tunable InfraRed Multi-component Acquisition system for plasma diagnostics", *Rev. Sci. Instr.* **71**, 3706-3710 (2000).

L. Mechold, J. Röpcke, X. Duten and A. Rousseau, „On the Hydrocarbon Chemistry in a H_2 Surface Wave Discharge Containing Methane", *Plasma Sources Sci. Technol.* **10**, 52-60 (2001).

J. Röpcke, L. Mechold, X. Duten and A. Rousseau, „A Time Resolved Laser Study of Hydrocarbon Chemistry in H_2-CH_4 Surface Wave Plasmas", *J. Phys. D: Appl. Phys.*, submitted (2001).

L. Mechold, J. Röpcke, D. Loffhagen and P.B. Davies, „Infrared Absorption Diagnostics and Chemical Modelling of Microwave Plasmas Containing Hydrocarbons", IVth International Workshop on Microwave Discharges: Fundamentals and Applications, in print (2001).